SECTION ON PHARMACOLOGY OF THE
INTERNATIONAL UNION OF PHYSIOLOGICAL SCIENCES
(SEPHAR)

PROCEEDINGS OF THE SECOND INTERNATIONAL PHARMACOLOGICAL MEETING

August 20—23, 1963

General Editor: Helena Rašková

Assistant General Editor: Jiří Vaněček

★

VOLUME 7

PROCEEDINGS OF THE FIRST INTERNATIONAL PHARMACOLOGICAL MEETING
STOCKHOLM, 22-25 AUGUST, 1961

Vol. 1 Part 1: *Plenary Session*
Part 2: *Pharmacological Control of Release of Hormones Including Antidiabetic Drugs*
Vol. 2 *Effects of Drugs on Synthesis and Mobilization of Lipids*
Vol. 3 *New Aspects of Cardiac Glycosides*
Vol. 4 *Drugs and Membranes*
Vol. 5 *Methods for the Study of Pharmacological Effects at Cellular and Subcellular Levels*
Vol. 6 *Metabolic Factors Controlling Duration of Drug Action*
Vol. 7 *Modern Concepts in the Relationship between Structure and Pharmacological Activity*
Vol. 8 *Pharmacological Analysis of Central Nervous Action*
Vol. 9 Part 1: *Bradykinin and Vaso-dilating Polypeptides*
Part 2: *Pharmacology of the Lung*
Vol. 10 *Abstracts*

PROCEEDINGS OF THE SECOND INTERNATIONAL PHARMACOLOGICAL MEETING
PRAGUE, 20-23 AUGUST, 1963

Vol. 1 *Pharmacology of Conditioning, Learning and Retention*
Vol. 2 *Biochemical and Neurophysiological Correlation of Centrally Acting Drugs*
Vol. 3 *Pharmacology of Cholinergic and Adrenergic Transmission*
Vol. 4 *Drugs and Enzymes*
Vol. 5 *Pharmacology of Cardiac Function*
Vol. 6 *Pharmacology of Smooth Muscle*
Vol. 7 *Pharmacology of Oriental Plants*
Vol. 8 *Evaluation of New Drugs in Man*
Vol. 9 *Recent Advances in the Pharmacology of Toxins*
Vol. 10 *Oxytocin, Vasopressin and their Structural Analogues*
Vol. 11 *Drugs and Respiration*

Pharmacology of Oriental Plants

Edited by

K. K. CHEN
Indiana University School of Medicine
Indianapolis, Indiana

B. MUKERJI
Chitharanjar National Cancer Research
Center, Calcutta

Assistant Editor

L. VOLICER
Institute of Pharmacology
Czechoslovak Academy of Sciences
Prague

CONGRESSUS PHARMACOLOGICUS II
· PRAGA 1963 ·

A Pergamon Press Book

THE MACMILLAN COMPANY
NEW YORK

THE MACMILLAN COMPANY
60 Fifth Avenue
New York 11, N.Y.

This book is distributed by
THE MACMILLAN COMPANY
pursuant to a special arrangement with
PERGAMON PRESS LIMITED
Oxford, England

This volume has been published jointly by PERGAMON PRESS
of Oxford and the CZECHOSLOVAK MEDICAL PRESS
of Prague

First edition 1965

Library of Congress Catalog Card Number 64-15320

Printed in Czechoslovakia

CONTENTS

LIST OF AUTHORS

Arora, C. K.	Department of Pharmacology, All India Institute of Medical Sciences, New Delhi, India
Arora, R. B.	Department of Pharmacology, All India Institute of Medical Sciences, New Delhi, India
Bein, H. J.	Research Laboratories of the Pharmaceutical Department of CIBA Limited, Basle, Switzerland
Brekhman, I. I.	Far-Eastern Section of the Siberian Branch of the Academy of Sciences of the U.S.S.R., Vladivostok, U.S.S.R.
Chen, K. K.	Indiana University School of Medicine, Indianapolis, Indiana, U.S.A.
Fu, F. Y.	Institute of Materia Medica, Chinese Academy of Medical Sciences, Peking, China
Ishii, Y.	Faculty of Pharmaceutical Sciences, University of Tokyo, Japan
Janků, I.	Institute of Pharmacology, Czechoslovak Academy of Sciences, Prague, Czechoslovakia
Jiřička, Z.	Institute of Pharmacology, Czechoslovak Academy of Sciences, Prague, Czechoslovakia
Lei, H. P.	Institute of Materia Medica, Chinese Academy of Medical Sciences, Peking, China
Motl, O.	Institute of Organic Chemistry and Biochemistry, Czechoslovak Academy of Sciences, Prague, Czechoslovakia
Mukerji, B.	Chitharanjar National Cancer Research Centre, Calcutta, India
Noble, R. L.	Cancer Research Centre, University of British Columbia, Vancouver, Canada
Petkov, V.	Department of Pharmacology, Postgraduate Medical Institute, Sofia, Bulgaria
Staneva-Stoicheva, D.	Department of Pharmacology, Postgraduate Medical Institute, Sofia, Bulgaria

TAKAGI, K.	Faculty of Pharmaceutical Sciences, University of Tokyo, Tokyo, Japan
VOLICER, L.	Institute of Pharmacology, Czechoslovak Academy of Sciences, Prague, Czechoslovakia
WATANABE, K.	Faculty of Pharmaceutical Sciences, University of Tokyo, Tokyo, Japan
ZHOU, J. H.	Institute of Materia Medica, Chinese Academy of Medical Sciences, Peking, China

PEPTIC ULCER-INHIBITING ACTIVITY OF LICORICE ROOT

K. TAKAGI, K. WATANABE and Y. ISHII

Faculty of Pharmaceutical Sciences, University of Tokyo, Bunkyo-Ku, Tokyo, Japan

CHINESE medicine, which had developed in ancient China, had been brought to Japan and had undergone a unique development here, before western medical science was introduced into the Meiji era. It was recently re-established as oriental medicine and has attained some success in the therapy of diseases which were very difficult to cure with western medicine.

The remedies used in oriental medicine consist of many plant products. According to western ideas of science, the biologically active ingredients must be isolated at first from the plant and then their pharmacological activities have to be studied. Such study presents serious difficulties for reasons described below.

Generally speaking, there exist many pharmacologically active ingredients in one plant product, some of them having a similar pharmacological activity, exerting a synergistic effect, some having antagonistic activity, and some others protecting chemically or physically the decomposition of the active principles. Moreover, in oriental medicine, use of a single product is very rare and several kinds of plant products are mixed for general use. Some prescriptions contain more than ten products. Various ingredients contained in one prescription exhibit a synergistic or antagonistic effect according to physiological conditions of patients. It is not a rare case that the same ingredient exhibits quite the opposite action when the associates of the combination are changed. Except alkaloids, which have such intense activity as ephedrine in *Ephedra*, or aconitine in *Aconitum*, no pharmacologically violent principles are found in prescriptions of oriental medicine. Their main ingredients exert so mild an activity that they cannot change normal conditions of living organisms, but they can exhibit a normalizing effect on abnormal conditions of patients. We must expect difficulties in demonstrating such a weak pharmacologic activity.

Even if an active principle could be isolated from a plant product, it might not represent all the therapeutic effects of that plant. At present we must confine ourselves to the study of oriental medicinal products only as a source of new drugs to be found after the scientific methods of western medicine.

We adopted licorice root at the first step of the study of crude plant products, because it is prescribed most frequently in oriental medicine. It is thought to have two actions: (1) a depressive, palliative action, and (2) a potentiating action to other drugs. It is used by itself for flavoring, pain relieving and as a detoxifying agent, and is used for throat pain. When it is combined with platycodon, action of the latter is potentiated. On the other hand, the anti-ulcer action of licorice extracts has not been recognized in the field of oriental medicine. The activity was discovered in 1946 by Dr Revers, who reported licorice to be effective in gastric ulcer. During the pharmacological study of anti-ulcer agents, we became aware of the fact that autonomic drugs, though used for duodenal ulcer, invariably give some autonomic side-effects and that their inhibitory action on stomach smooth muscle rather give ill effects on gastric ulcer. Because there seemed to be no synthetic drugs which really inhibit humoral gastric secretion, and do not suppress gastric motility, at the beginning of this research the anti-ulcer activity of licorice root was investigated. The pharmacological activity of licorice, which might be regarded as having some relation to anti-ulcer action, was anti-inflammatory action (Cornforth and Long, 1957) and inhibitory action of ulcers of Shay rats (Desmarez, 1953). The glycoside glycyrrhizin or its aglycon glycyrrhetinic acid has anti-inflammatory action, but no ulcer-protecting action to Shay rats (Nomura, 1959). The fraction, which is devoid of glycyrrhizin, definitely exerts anti-ulcer activity. We tried to confirm the results of previous reports and further the mode of action of licorice to inhibit gastric acid secretion.

Extraction of the active principles of licorice root (Fig. 1)

Methanol extract (F_M). One kg of powdered licorice, *Glycyrrhiza glabra* L. var. *glandulifera* Regel et Herder, was shaken for 10 hr with 5 l. of methanol at room temperature and the insoluble residue was treated similarly twice more. The combined filtrate was concentrated and dried *in vacuo*. The residue was a brownish-black resinous mass. The yield was 20 per cent.

Purified methanol extract (F_M'). The combined methanol solution obtained as above was passed through a column of 500 g of Tolite MG-S, and the filtrate was concentrated *in vacuo*. The residue was a brownish powder. The yield was about 15 per cent.

Water extract (F_W). Water (5 l.) was added to 1 kg of powdered licorice and heated for 8 hr at 100°C. The filtrate was concentrated *in vacuo*. The yield of dried residue was 17 per cent.

Effect of the extracts on experimental peptic ulcer

Shay rat. After 48 hr fasting, the pylori of male rats were ligated and each

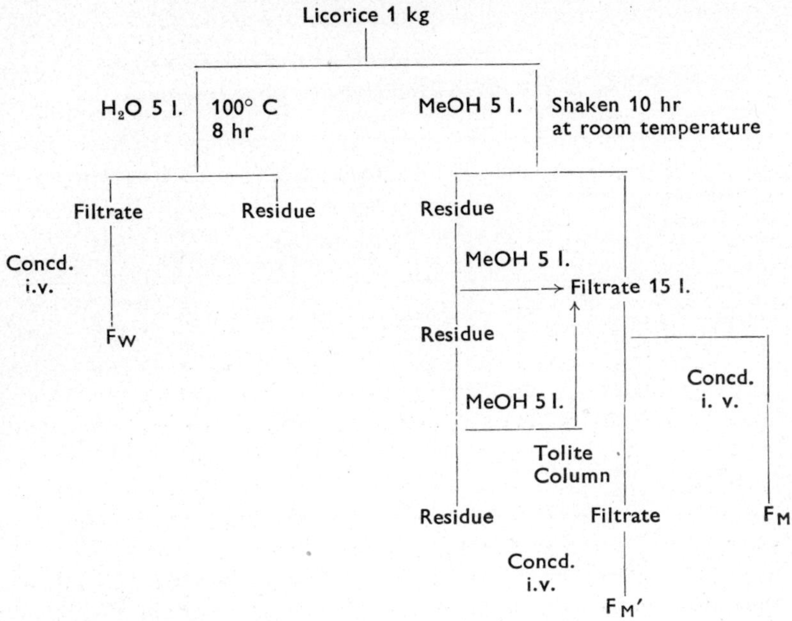

FIG. 1. Extractions of licorice root.

fraction suspended in 0·5 per cent CMC solution, administered intraperitoneally immediately after the ligation in a dose of 0·5 ml/100 g body weight, and the control animals were given 0·5 ml of 0·5 per cent CMC solution per 100 g. After 19 hr, the animals were sacrificed, the stomachs removed, the gastric contents tested for acidity, and the damage of gastric mucosa estimated. The ulcer index was given by the ulcer diameter as shown in Table 1. The protective action of the methanol extract F_M from Shay ulcer was superior to the water extract F_W in all the properties tested, such as incidence of ulcer, ulcer index, and volume and pH of gastric content. It was then confirmed that F_M was able to inhibit Shay ulcer of rats by oral route as shown in Table 2. Glycyrrhizin does not inhibit gastric secretion, but the ulcer index of Shay rats is very low. These facts suggest the existence of some mechanisms of ulcer protection other than gastric acid inhibition.

Stress ulcer. A male rat was fixed on a board and immersed up to the breast in water of 25°C for 20 hr (Takagi and Watanabe, 1962). Haemorrhage and erosion developed in almost all the animals on the mucosa of the glandular portion of the stomach. This stress ulcer can be depressed by cholinolytics and to some extent by CNS depressants. Pepsin inhibitors such as heparin and similar polysaccharide sulfate esters can indeed inhibit Shay ulcer but not protect completely the stress ulcer of rats (Namekata,

TABLE 1

EFFECT OF EACH FRACTION OF LICORICE EXTRACT ON SHAY RAT ULCER ADMINISTERED INTRAPERITONEALLY

Treatment (mg/kg)		No. of animals	Ulcer		Gastric secretion		
			Incidence (%)	Index*	Volume ml/100 g	pH	Free acid (mEq/l.)
Control		77	93·6	24·72	6·35	2·2	28·0
F_M	500	33	0	0	1·39	5·2	4·0
F_M	250	5	60	7·92	4·55	1·9	22·5
F_M'	500	18	0	0	0·82	3·9	3·8
F_M'	250	17	11·8	5·88	2·66	4·2	20·3
F_W	500	10	40	1·98	4·42	3·6	22·3
Glycyrrhizin	500	9	77·8	0·33	8·01	1·3	60·9

*Ulcer index is given as follows:

Diameter of ulcer (mm)	0—0·5	0·5—2	2—4	4—6	6—10	10<
Ulcer index	0·2	1·0	2·0	3·0	4·0	10·0

Perforated stomach is given index fifty in total.

TABLE 2

EFFECT OF F_M ON SHAY RAT ULCER GIVEN ORALLY 2 g/kg BEFORE THE LIGATION AND 1 g/kg INTRADUODENALLY AFTER THE LIGATION

Treatment	No. of animals	Ulcer		Gastric secretion		
		Incidence (%)	Index	Volume ml/100 g	pH	Free acid (mEq/l.)
Control	5	100	50	All animals perforated		
F_M	5	80	0·7	4·23	1·5	55·5

1962). Two licorice extracts, F_M and F_W, inhibited stress ulcer but not completely, as indicated in Table 3.

TABLE 3

EFFECT OF THE LICORICE EXTRACTS ON THE STRESS ULCER OF RATS, WHICH IS RESTRAINED AND IMMERSED IN WATER OF 25°C FOR 20 HR

Treatment	No. of animals	Route of administration	Dose (mg/kg)	Ulcer		pH
				Incidence (%)	Index	
Control	6	s. c.	500	83·3	9·0	3·9
F_W	6	s. c.	500	66·7	1·9	4·2
F_M	6	s. c.	500	33·3	2·5	4·4
F_M	7	i. p.	500	57·2	4·4	4·3
Control	10	i. p.	500	90·0	11·5	3·8

Reserpine ulcer. Male rats were injected subcutaneously 5 mg/kg of reserpine and kept from food for 20 hr. F_M (500 mg/kg) was given intraperitoneally simultaneously with reserpine. Ulcer incidence and ulcer index were not lowered by F_M (Table 4).

TABLE 4

EFFECT OF F_M (500 mg/kg i.p.) ON RESERPINE (5 mg/kg s.c.) ULCER OF RATS

Treatment	No. of animals	Death	Ulcer incidence (%)	Index
Control	10	2	100	11·8
F_M	10	3	100	11·4

Histamine ulcer. Male guinea-pigs were given 10 mg/kg of histamine once a day for 3 weeks, suspended in 10 mg/ml in a mixture of liquid paraffin and bees-wax (4 : 1) (Hay, 1942). Just before the administration of histamine, 2 mg/kg of diphenhydramine was injected subcutaneously to prevent histamine shock. F_M (250 mg/kg) was given orally once a day for 3 weeks. During the experiment some of the animals died before the end of the experiment and their stomachs were examined by autopsy. Because the lethality without ulcer was high in the F_M-treated group, results in Table 5 were rather inaccurate, but the incidence of ulcer in the F_M-treated group was definitely lower than that of the control group.

TABLE 5

EFFECT OF F_M ON HISTAMINE ULCER OF GUINEA-PIGS

	Ulcer incidence (%)	Death (%)
Control	13/15 86·7	11/15 73·3
F_M	6/15 40·0	13/15 86·7

Histamine 10 mg/kg/day i.m., diphenhydramine HCl s.c. 2 mg/kg, F_M 250 mg/kg/day p.o. Given for three weeks.

Effect on pepsin activity

Pepsin activity of rat stomach juice, which was collected 19 hr after the ligation of pylorus of a rat, was estimated by the hemoglobin-Folin colorimetric method (Anson, 1938). $F_{M'}$ (250 mg/kg), administered intraperitoneally immediately after the ligation, exhibits inhibition of 33·4 per cent of pepsin activity of rat stomach juice, which is significant at 1 per cent probability.

$F_M{}'$ 2·8 ± 0·22 mg as tyrosine/ml
Control 4·2 ± 0·56 mg as tyrosine/ml

Carbonic anhydrase inhibition by $F_{M'}$

Hemolysed total blood of rabbit was used as the enzyme preparation and sodium bicarbonate as substrate. $F_{M'}$ did not show any inhibitory action on the enzyme at 10^{-4} g/ml and, on the contrary, acetazolamide showed definite inhibition at 10^{-7} g/ml.

Anti-inflammatory action

Both glycyrrhizin and the glycyrrhizin-free fraction of licorice were reported to have anti-inflammatory action on formalin edema of rat's paw and on cotton-pellet granuloma of rats (Cornforth and Long, 1957; Finney and Somers, 1958). Inhibitory action of F_M, F_W, and glycyrrhizin on dextran edema of rat's paw is indicated in Fig. 2.

Antispasmodic and anti-acetylcholine action

F_M has weak antispasmodic action on isolated ileums of guinea-pigs as shown in Fig. 3. It has a papaverine-like musculotropic effect, and antihistamine and anti-acetylcholine activity are feeble. The activity is attributed to the aglycons of glycosides, as liquiritigenin and isoliquiritigenin (Shibata and Harada, 1960), but because papaverine itself has no anti-ulcer activity on Shay rats, the antispasmodic action does not seem to contribute to anti-ulcer activity of licorice.

The spontaneous movements of rat's stomach and the contraction, which was temporarily elicited by methacholine, of anesthetized rats remained unchanged with F_M, although they were completely depressed by atropine sulfate, and the temporary hypotensive activity of ACh on the blood pressure of dog femoral artery was not blocked by F_M. All these results indicate that F_M has negligible atropine-like anti-acetylcholine activity.

Other actions

Prolonging action of hexobarbital sodium hypnosis of mice and local anesthetic action on guinea-pig cornea and on rat sciatic nerve of F_M were feeble. Blood pressure and respiration of urethan-anesthetized dogs were little influenced.

Acute toxicity

LD_{50} on mice was estimated 72 hr after intraperitoneal injection of the extracts.

$$F_M \quad LD_{50} = 1{\cdot}33 \text{ g/kg } (1{\cdot}21 \sim 1{\cdot}46 \text{ g/kg}, P = 0{\cdot}05)$$
$$F_W \quad LD_{50} = 2{\cdot}52 \text{ g/kg } (2{\cdot}04 \sim 3{\cdot}10 \text{ g/kg}, P = 0{\cdot}05)$$

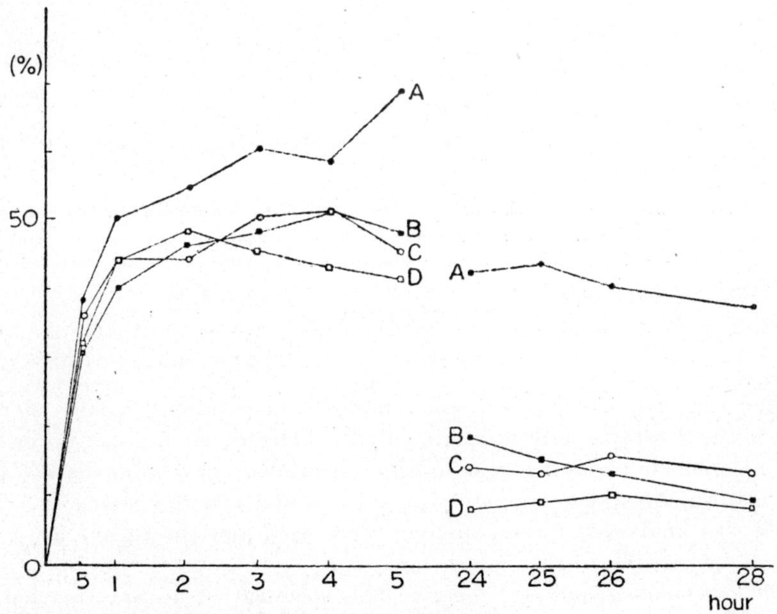

FIG. 2. Anti-inflammatory action of licorice extracts on dextran edema of rat's paw. Ordinate is percent increase of the paw injected 6 per cent dextran subcutaneously. Abscissa, time after the injection of dextran. A: control, B: F_M' 250 mg/kg i. p., C: Glycyrrhizin 250 mg/kg i. p., D: F_W 250 mg/kg i.p.

Effect of the methanol extract on gastric secretion

From the preceding experiments, licorice seems to contain some substances which inhibit gastric secretion and have very weak anti-acetylcholine activity. These facts suggest further the possibility that licorice suppresses secretion through the inhibition of a humoral mechanism.

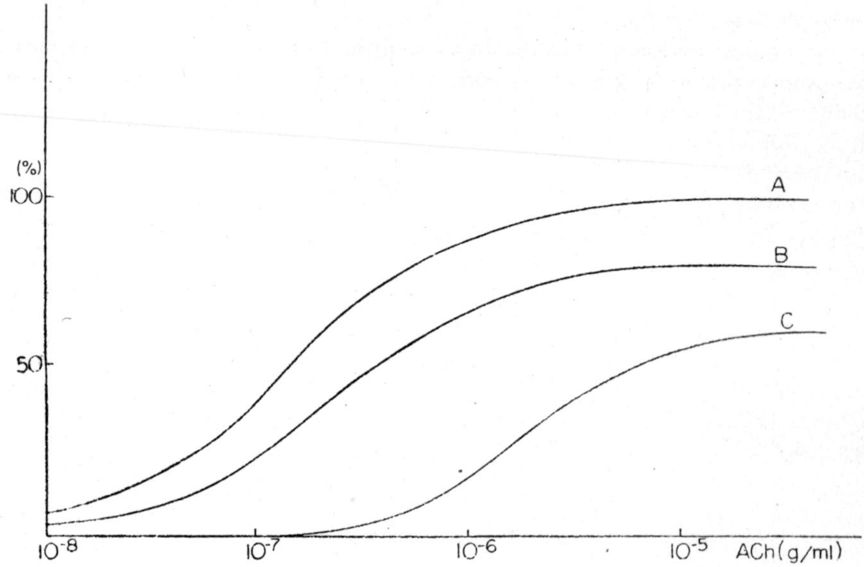

FIG. 3. Antispasmodic action of F_M on excised ileums of guinea-pigs. Ordinate shows percent contraction of the ileums to the maximum contraction and abscissa is concentration of acetylcholine in the bath solution. A is the dose-response curve of ACh, B is that with F_M 10^{-5} g/ml, and C is that with F_M 10^{-4} g/ml.

Shay rats. F_M (250 mg/kg) was administered intraperitoneally 30 min before the ligation, and histamine hydrochloride (20 mg/kg) or methacholine chloride (1 mg/kg) was injected intramuscularly immediately after the ligation. After 4 hr, the stomach was excised and the gastric juice collected was analysed. Twelve animals were used and the results are given in Table 6 and Fig. 4.

Rat with acute fistula on stomach. The stomach of an anesthetized rat was ligated at the pylorus and a double-tubed polyethylene cannula was inserted and fixed at the forestomach. Every hour the stomach was cleansed twice with 3 ml of physiological saline and the combined washing solution was titrated with 0·1 N-NaOH. Histamine hydrochloride (10 mg/kg) was

TABLE 6
EFFECT OF F_M ON GASTRIC ACID SECRETION (ml/100 g) AND pH OF GASTRIC JUICE

	Stimulant	—		Histamine 20 mg/kg i.m.		Methacholine 1 mg/kg i.m.	
	Inhibitor	—	F_M	—	F_M	—	F_M
Secretion (ml/100 g)	Mean	0·95	0·63*	1·50	0·94*	1·97	0·95†
	S.E.	0·05	0·10	0·13	0·08	0·21	0·10
pH	Mean	3·01	3·93*	1·45	1·83*	1·98	3·70†
	S.E.	0·19	0·36	0·09	0·14	0·22	0·54

* Significant at $P = 0{\cdot}05$. † Significant at $P = 0{\cdot}01$.

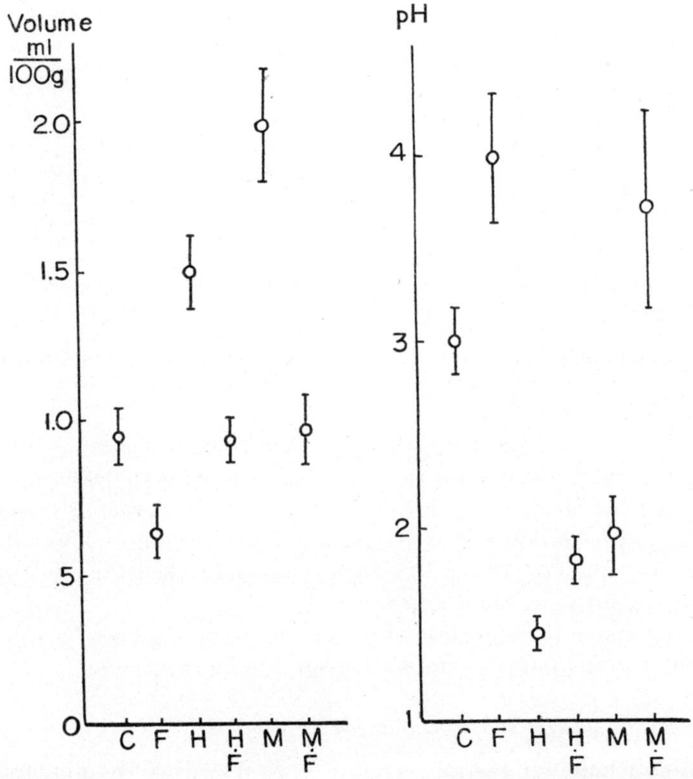

FIG. 4. Effect of F_M on gastric acid secretion and pH of gastric juice of pylorus ligated rats. *C* is control, *F* is F_M 250 mg/kg i.p., *H* is histamine HCl 20 mg/kg i.m., and *M* is methacholine 1 mg/kg i.m. Each point is the mean of twelve animals with standard error.

injected subcutaneously. $F_M{}'$ (200 mg/kg) intraperitoneally inhibited the histamine-induced gastric secretion as indicated in Fig. 5.

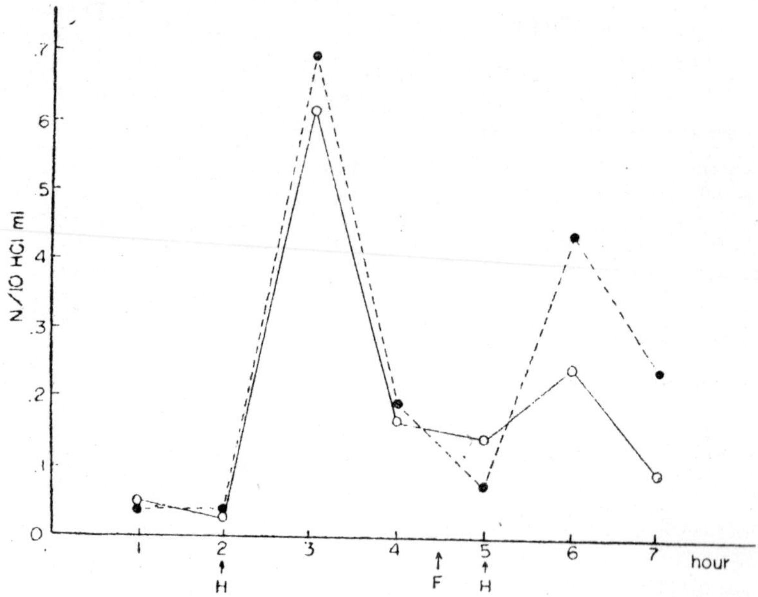

FIG. 5. Inhibitory action of $F_M{}'$ on gastric secretion by histamine, collected every hour in the pylorus ligated rats with acute fistula at forestomach. The control group (● — — — ●) received at *H* histamine HCl 10 mg/kg s.c. and the treated group (o——o) received at *F* $F_M{}'$ 200 mg/kg i.p. Each point is the mean of six animals.

Schild's perfusion method of rat stomach. According to the report of Schild (Ghosh and Schild, 1958), a rat stomach was perfused with 0·00025 N-NaOH solution and the change of pH was recorded. The intravenous injection of histamine hydrochloride 100 to 200 μg per animal definitely lowered pH of the perfusate and F_M (10 mg/kg) intravenously blocked the effect of histamine 200 μg for more than 2 hr (Fig. 6).

From the above three experiments, inhibition by F_M may be nonspecific and an interaction with gastrin mechanism can be supposed.

Effect of F_M on gastric secretion induced by peptone

Increase of humoral gastric secretion is produced by the application of peptone on the antrum of a stomach. We studied the effect of F_M on peptone-induced gastric secretion of rats. The stomach of a rat, anesthetized by pentobarbital, was exposed and separated by small forceps between pylorus and glandular stomach as indicated in Fig. 7. A cannula was fixed in the forestomach and it was cleansed every hour with physiological saline. Ten

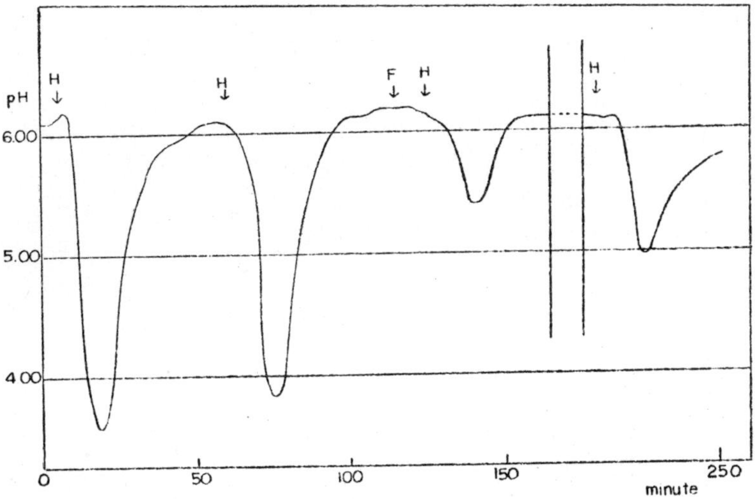

FIG. 6. Changes of pH of the alkaline perfusate (0·00025 N-NaOH) of rat stomach by histamine and F_M. *H* is histamine HCl 200 μg i.v. and *F* is F_M 10 mg/kg i.v.

per cent peptone solution 0·2 ml was injected every 2 hr into the antrum and the second and third injections secreted more gastric acid than the first

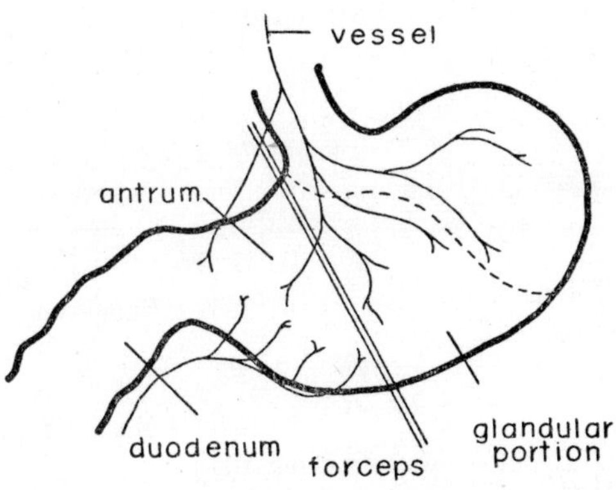

FIG. 7. Separation of rat's stomach by forceps between pylorus and glandular stomach.

peptone application. F_M (100 mg/kg) intraperitoneally completely suppressed the secretion (Fig. 8).

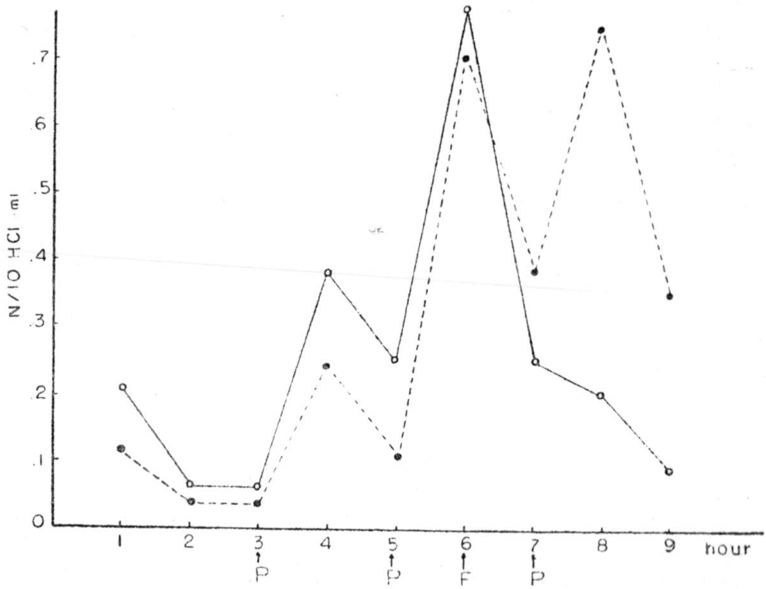

FIG. 8. Effect of F_M on peptone-induced gastric acid secretion of rats. The control group (● — — — ●) received at *P* 10% peptone solution 0·2 ml and the gastric juice is collected every hour from the forestomach. The treated group (o——o) received at *F* F_M 100 mg/kg i.p. Each point is a mean of twenty animals.

The peptone-induced gastric secretion was inhibited by atropine sulfate 5 mg/kg and hexamethonium 5 mg/kg intraperitoneally, and by the application of dibucaine hydrochloride 2 per cent solution on pyloric mucosa as shown in Table 7.

TABLE 7

EFFECT OF SOME DRUGS ON PEPTONE-INDUCED GASTRIC ACID SECRETION

Stimulant in pylorus	Inhibitor	Dose	No. of test	0·1 N-HCl (ml)
Saline	—	—	33	0·07 ± 0·02
Peptone	—	—	33	0·49 ± 0·05
Peptone	Atropine sulfate	5 mg/kg i.p.	9	0·04 ± 0·01
	Hexamethonium	5 mg/kg i.p.	7	0·13 ± 0·03
	Dibucaine HCl	2% to antrum	8	0·13 ± 0·03

These facts indicate the mediation of the gastric mechanism in peptone-induced gastric secretion and the action of F_M might be realized through affecting the gastrin mechanism.

Effect of F_M on gastric secretion through gastrin

Gastrin preparation. According to Gregory, the pyloric antrum of three rats was extracted with 6·0 ml of TCA–acetone solution and a crude gastrin was prepared (Gregory and Tracy, 1961). The preparation was 3 ml of aqueous solution and each 0·25 ml was injected intravenously. The acetylcholine content of this preparation was tested on the eserinized rectus abdominis of a frog and proved to be below $3{\cdot}3 \times 10^{-7}$ g/ml. This preparation was also estimated to contain about 10^{-6} g/ml of histamine on excised ileum of guinea-pigs. Because the gastric secretion of rats was not influenced by 0·2 μg of acetylcholine and 20 μg of histamine per animal, they might not disturb the exact determination of gastrin effect.

Similar extracts from glandular stomach, forestomach, and duodenum of rats exhibited feeble or no stimulation of gastric secretion as shown in Fig. 9.

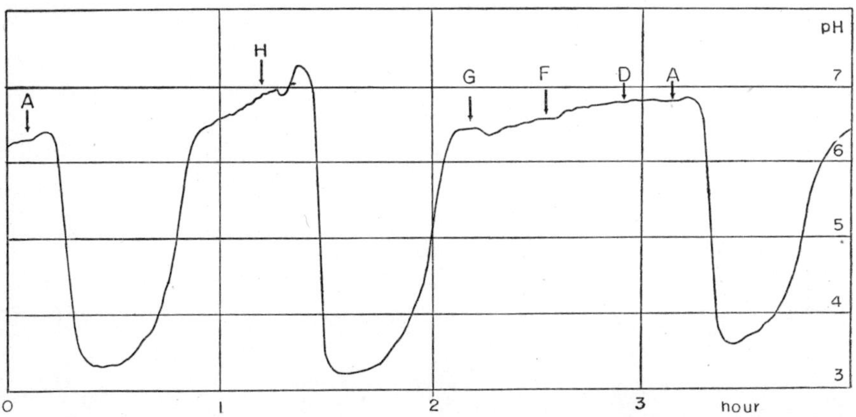

FIG. 9. Stimulation of gastric acid secretion by TCA-acetone extract of various parts of gastrointestinal tract. pH change of the perfusate of rat stomach is recorded. *A* is half the extract of pyloric antrum of a rat, and *G*, *F*, and *D* are the extracts of glandular stomach, forestomach and duodenum of a rat. *H* is histamine HCl 1 mg/kg. All the test solutions are injected intravenously.

Effect of F_M on gastric secretion by the crude gastrin preparation. Activity of the gastrin preparation and the antagonism of F_M to gastrin were assayed on perfusion of rat stomach. The results are shown in Table 8 and Fig. 10. F_M exerts some inhibition on the secretion by gastrin, but it is not so significant as the inhibition of the histamine-induced gastric secretion by F_M.

TABLE 8

LOWERING OF pH OF THE PERFUSATE OF RAT STOMACH BY HISTAMINE AND GASTRIN, BEFORE AND AFTER THE ADMINISTRATION OF F_M

Animal	Before F_M		After F_M	
	Histamine	Gastrin	Histamine	Gastrin
1	2·7	2·8	1·6	2·2
2	3·9	3·7	3·0	3·3
3	3·5	0·9	3·3	3·1
4	1·6	1·2	0·8	0·5
5	2·0	1·0	1·2	0·3
6	3·3	2·1	1·8	2·0
Mean	2·84	1·95	1·95	1·90
Per cent	100	100	68·6	97·5

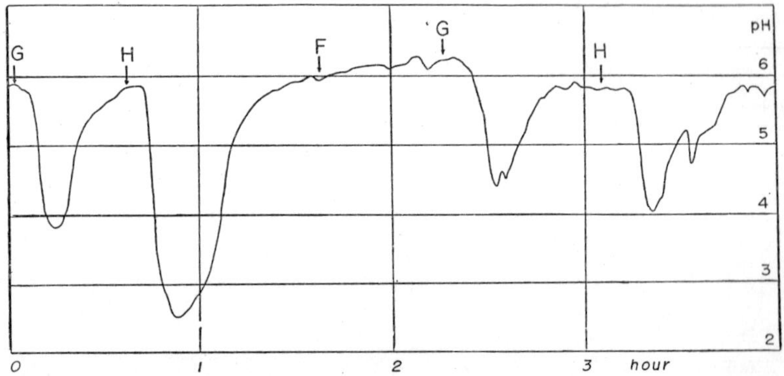

FIG. 10. Effect of F_M on gastrin-induced gastric secretion of a rat. pH change of the perfusate of rat stomach is recorded. *G* is a quarter of the crude gastrin preparation from a rat, *H* is 1 mg/kg of histamine HCl, and F is F_M 250 mg/kg. *G* and *H* are injected intravenously and *F* intraperitoneally.

CONCLUSION

The best-known ingredient of licorice is glycyrrhizin, which has anti-inflammatory action and salt-retaining action like mineralocorticoids.

Although the anti-inflammatory action may contribute to the anti-ulcer activity of licorice, it does not represent the whole activity of the plant. Some ingredients other than glycyrrhizin are proved to inhibit gastric secretion, which has been increased by methacholine, histamine, and by the application of peptone on the antrum of stomach. These activities are different from those of atropine, which inhibits the secretory action of methacholine and peptone, but does not inhibit that of histamine, and suggests the interaction of F_M with the humoral mechanism of gastric secretion. On the other hand, F_M has low activity to antagonize the gastrin preparation and consequently we suppose that F_M should inhibit mainly the liberation of gastrin from pyloric mucosa. The fact that F_M has little anti-acetylcholine action and that it does not inhibit spontaneous motility of rat stomach *in situ*, although atropine completely inhibits the motility, emphasizes the utility of the licorice extract for treating gastric ulcer without cholinergic ill effects.

REFERENCES

ANSON, N. L. *J. Gen. Physiol.* **22**, 79 (1938).

CORNFORTH, J. W. and LONG, D. A. *Brit. Med. J.* **1**, 340 (1957).

DESMAREZ, J. J. *C.R. Soc. Biol.* **150**, 1022 (1953).

FINNEY, R. S. H. and SOMERS, G. G. *J. Pharm. Pharmacol.* **10**, 613 (1958).

GHOSH, M. N. and SCHILD, H. O. *Brit. J. Pharmacol.* **13**, 54 (1958).

GREGORY, R. A. and TRACY, H. J. *J. Physiol.* **156**, 523 (1961).

HAY, L. J., VARCO, R. L., CODE, C. F. and WAGENSTEEN, O. H. *Surg. Gynec. Obstet.* **75**, 170 (1942).

MELDRUM, N. U. and ROUGHTON, F. J. W. *J. Physiol.* **80**, 113, 143 (1933).

NAMEKATA, M. *Chem. Pharm. Bull.* **10**, 177 (1962).

NOMURA, H. *Fukuoka Acta Medica* **50**, 357 (1959).

REVERS, F. E. *Ned. Tijdschr. Geneesk.* **90**, 135 (1946).

SHIBATA, S. and HARADA, M. *Yakugakuzasshi* **80**, 620 (1960).

TAKAGI, K. and WATANABE, K. *Folia Pharmacol. Japon* **58**, 4, 144§ (1962).

RECENT ADVANCES IN THE PHARMACOLOGICAL RESEARCH OF MEDICINAL PLANTS IN THE PEOPLE'S REPUBLIC OF CHINA

J. H. Zhou, F. Y. Fu and H. P. Lei

Institute of Materia Medica, Academia Sinica, Shanghai, and Institute of Materia Medica, Chinese Academy of Medical Sciences, Peking, China

The history of the use of medicinal plants in China is one without interruption for thousands of years. During the last century, owing to the aggression of imperialism and misrule of the feudal governments of the old China, Chinese medicine and folklore drugs have been seriously discriminated and neglected; research was almost at a complete standstill. This was most evident in the regime of the Kuomingtang reactionary clique.

Since 1949, with the founding of the People's Republic of China, under the remarkable leadership of the Communist Party of China and Chairman Mao Tze-tung, great strides in pharmacological research and pharmaceutical production have been taking place with a speed unprecedented in the history of modern medicine. These achievements are primarily due to the policy of emphasizing the closer unity and cooperation between the schools of traditional medicine and modern medicine for the well-being of our national medical sciences and healing arts. New schools and research institutes have been established in all parts of our vast territory and research centers created in the leading cities. All measures and efforts are directed towards the promotion of the health of the broad masses of the 600 million working people.

Since 1958 the banner of the great leap forward in our socialist construction has encouraged our people with more confidence than ever, even in the face of temporary difficulties in crop-harvesting of 1959—1961, in the endeavour to greater heights in the fields of medicine and pharmacology.

New drugs have been rediscovered from our ancient heritage and these are prepared in a purer chemical state.

Many biologically active alkaloids, such as liensinine from lotus seeds, glucosides, lactones, such as yejuhua lactone from *Chrysanthemum indicum*, steroids and other principles, including one new amino acid cucurbitine from the pumpkin seeds have been recently isolated and their structural configurations partially or completely worked out. These preliminary achievements which were carried out in a short span of less than ten years are testimonials speaking for the tremendous possibilities awaiting further work along the line of investigating the ancient Chinese medical legacy by modern scientific technology and dialectical ideology.

Space does not allow for even a passing account of the various aspects of pharmacological research on medicinal plants in China. Only a few highlights will be presented here.

1. THE PLACE OF PHARMACOLOGICAL RESEARCH IN SOCIALIST CHINA

The main objectives of our medical and health program are: prevention and cure of diseases endangering the lives and health of the broad masses of the people; promotion of labor hygiene and prevention of occupational diseases; development of Chinese traditional medicine; higher scientific and technical standards in pharmaceutical industry; and promotion of social hygiene, higher nutritional standards for the people and a nation-wide program of athletic sports.

The Chinese pharmacologists and pharmaco-chemists regard it as their primary task to observe the aforementioned objectives in conducting their research projects. In the short decade following the founding of the People's Republic of China, most of the infectious diseases once rampant in our cities and villages such as smallpox, plague, cholera, syphilis, gonorrhea, kala-azar, leprosy are now basically eradicated. Tuberculosis has been greatly reduced. Schistosomiasis, once prevailing and threatening the lives of over 100 million people in Central and South China, is now practically arrested. Obviously, our attention is now directed mainly to drugs regulating the functional activity of the nervous system and neurohormones, antihypertensive drugs, cardiovascular drugs and anticancer agents. Pharmaceutical plants have been built and an independent pharmaceutical industry has been created on the principle of self-reliance to provide for most of our needs and is able to a certain extent in supplying some of our friendly countries.

China has different climatic and geographical characteristics. Owing to the variabilities of ecological conditions, the flora and natural resources of medicinal plants in China are abundant. Recent survey shows that out of 260 families, 2014 genus and 25,000 species of seed plants native to mainland China, about 3500 species are known to possess medicinal value.

2. DRUGS REGULATING THE FUNCTIONAL ACTIVITY OF THE NERVOUS SYSTEM AND NEUROHORMONES

Recent experimental studies on *zhenshen* (*Panax ginseng*) have added further positive evidence in support of the thesis that *Panax ginseng* promotes the defensive mechanism to stress, such as high environmental temperature, experimental febrile state, acute hemorrhagic shock, asphyxial state, histamine shock. During some of these stress responses, rats receiving *Panax ginseng* showed significantly reduced fall of the ascorbic acid content of the adrenal glands. It appears to be capable of lessening the edema at the ankle joint due to the injection of egg-white or dextran. It exerts certain specific effects on carbohydrate metabolism, such as partial control of hyperglycemia in alloxan diabetes. It has direct hemodynamic action on the cardiovascular system with an increase in the force of cardiac contraction, especially during acute circulatory failure.

Yen-hu-suo (*Corydalis ambigua* Cham. et Sch.) has been reputed as an analgesic in Chinese medicine. One of the many alkaloids isolated is Corydalis B, identified as tetrahydropalmatine (THP). Extensive laboratory and clinical investigations proved its usefulness in alleviating pain and anxiety in a variety of clinical conditions.

Tetrahydroberberine (THB), chemically closely related, is more tranquillizing and less analgesic. Palmatine is now extracted from another plant, Huang-teng (*Fibraurea tinctoria* L.), its high content of palmatine making the commercial preparation of tetrahydropalmatine readily available.

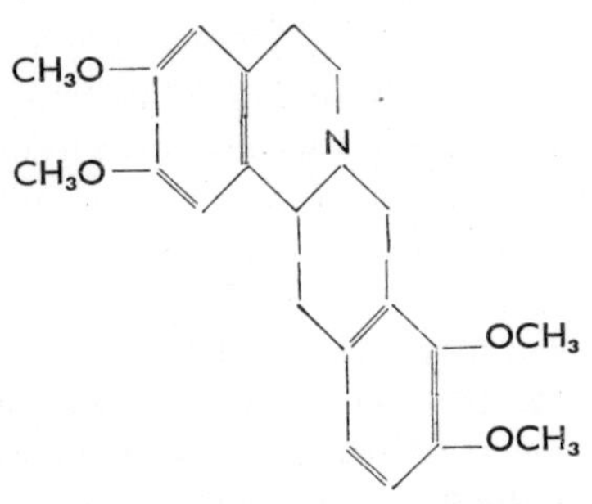

Tetrahydropalmatine

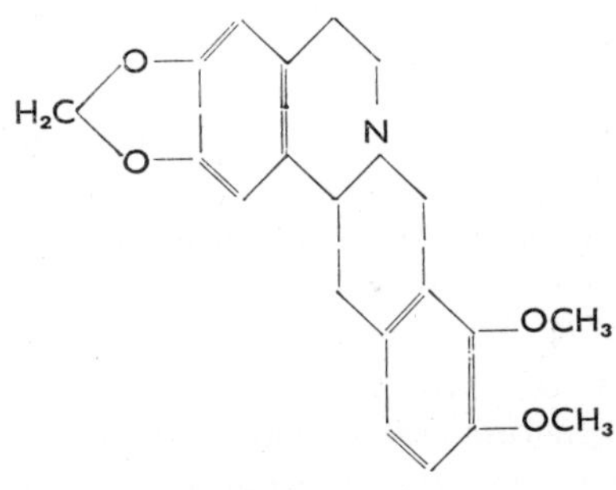

Tetrahydroberberine

Tetrahydropalmatine has a wide range of actions, as revealed from EEG studies, stereotaxic stimulation of different sites of the brain stem, *encéphalé isolé* preparation and intraventricular injection experiments. It increases also the ACTH secretory activity.

The riddle of the famous ancient surgical narcotic mixture used by the Chinese surgeon Huatuo in the second century was partially solved by the recent discovery of the marked potentiation of the analgesic action of rhododendron and aconitine by scopolamine. It may be pointed out that andomedotoxin, formerly believed to be the active principle of rhododendron, was found to possess a low degree of analgesic action and was high intoxicity. The analgesic agent was found in the fruit of the plant. The potency of the crude extract was half that of an equal dose of crude opium powder.

Many medicinal plants have been extracted and screened for analgesic action, especially for the use in alleviating arthritic inflammatory reactions. The methods adopted were usually the Woolfe–MacDonald heat radiation threshold test for analgesia, the Bacchus formaldehyde test, and the Winder egg-white test in albino rat for antiphlogistic action.

The following substances were noted for their positive effects:

(a) Craibiodendrin ($C_{21}H_{29}O_6$) isolated from gou-jiao-cao (*Crabiodendron stellatum* W. et M.).

(b) Gentianine and gentianidine from Qin-jiao (*Gentiana macrophylla*).

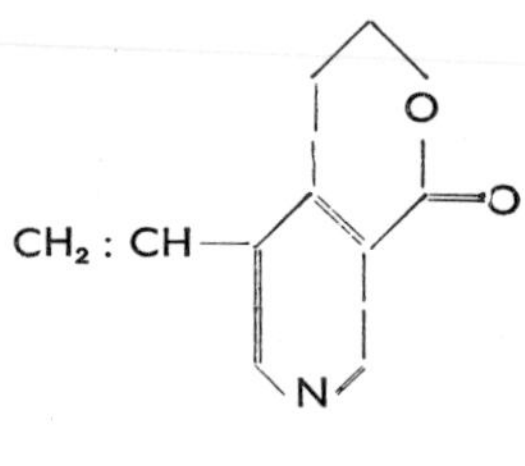

Gentianine

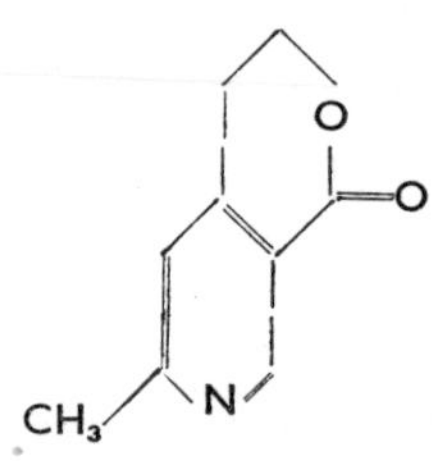

Gentianidine

(c) The total alkaloids from the mixture of two plants, Gui-zhen-cao (*Bidens birnata* L.) and Chou-wu-tong (*Clerodendron trichotomum* Thunb.).

These substances appeared to have similar mode of action in reducing the inflammatory arthritic swelling. They may act by way of pituitary–adrenal pathway.

It is of interest to note that vanillin was found to be the active anticonvulsive agent in Tien-ma (*Gastrodia elata* Bl.). This was demonstrated in experimental epilepsy in guinea-pigs, electroshock convulsion in rats, and metrazol convulsion in mice. Epileptic discharge in EEG was reduced by vanillin.

These few examples suffice to illustrate the fact that the practical clinical experiences in Chinese traditional medicine may be confirmed and extended by modern pharmacological investigations, provided proper methods are chosen.

3. ANTIHYPERTENSIVE DRUGS

Much of our effort in the last few years has been devoted to the search for and investigation of *Rauwolfia* species in our country. Thanks to the close collaboration of the botanical workers *Rauwolfia verticillata* L. Bail. was found in abundance in South and Southwest China. Other species were also discovered: *R. serpentina*, *R. parakensis*, *R. tialoshanensis*, *R. yunnanensis*, *R. latifrans* and others.

The Chinese and Indian species are more or less similar in their alkaloidal content and pharmacological actions. There are, however, certain differences. The total alkaloids of the Chinese species (*R. verticillata*) appear to be somewhat more potent in hypotensive activity, as tested on hypertensive dogs and rats, and produce less side-effects. This is borne out in the clinical reports.

Besides the alkaloids found in *R. serpentina*, new alkaloids were discovered in *R. verticillata*, namely *Rauwolfia* A ($C_{25}H_{28}N_2O_2$), samatine ($C_{20}H_{24}O_3N_2 \cdot 2{\cdot}5\ H_2O$), and samatinine ($C_{20}H_{25}O_2N_2^+ \cdot X^-$). These substances are weak in central sedative action and possess more marked peripheral action, such as adrenolytic action.

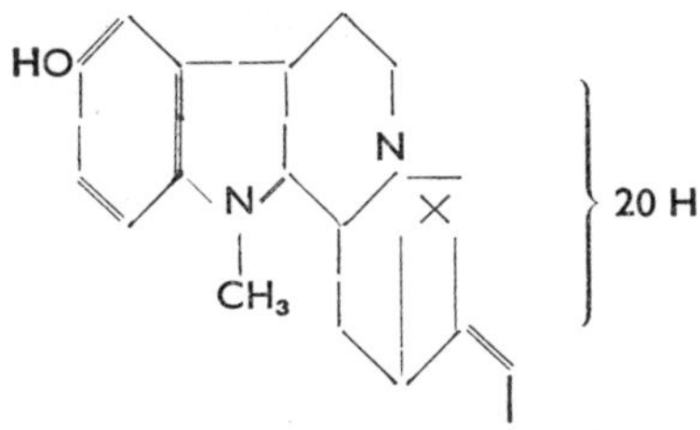

Samatine

Systematic studies were conducted on the hemodynamic mechanism of these alkaloids. Suffice to say, owing to the remarkable coordination among the different groups of scientists, clinicians and pharmaceutical workers, commercial preparations of Chinese *Rauwolfia* soon became available to supply our domestic market.

Other sources of hypotensive drugs among the medicinal plants were screened both in normotensive and hypertensive animals. Various laboratories have routinely employed renal ischemic hypertensive rats as well as the Goldblatt dogs for this purpose. A large number of the medicinal plants investigated in the past few years showed impressive antihypertensive effect both in the degree and duration of action. The following agents are of significance.

(a) Liensinine, from the embryo of lotus seed (*Nelumbo nucifera* Gaertner). It is a tertiary amine, resembling *d*-tubocurarine in chemical structure with the absence of an ether oxygen linkage. The hypotensive action is greatly increased on quarternization.

(b) Peonol, the volatile phenolic substance from the bark of peony (*Paeonia suffruticos* Andr.). Studies on synthetic peonol and derivatives have shown that the ketonic linkage was essential and the removal of the hydroxyl radical resulted in partial loss of activity. It was suggested that peonol probably exists in the form of water-soluble glucoside which is responsible for its hypotensive action.

(c) Tetrandrine and demethyltetrandrine from Hanfangji (*Stephania tetrandras* Moore). These alkaloids were found to possess a variety of pharmacological actions, of which the hypotensive action is of particular interest. As a result of vasodilatation of visceral organs through both the direct action on the vascular structure and predominantly the central vasomotor reflex mechanism, tetrandrine exerts marked and sustained hypotensive action. It has also moderate analgesic action of central origin, potentiated by diphenylhydramine, and anti-inflammatory action as well as an inhibitory effect on *Entameba histolytica*.

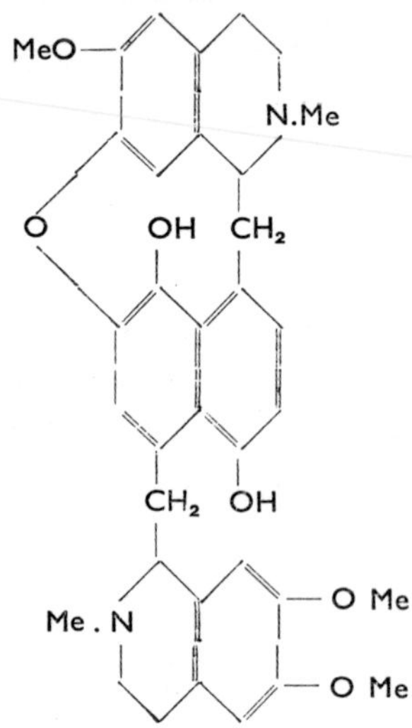

Liensinine

(d) Yejuhua (*Chrysanthemum indicum* L.) was discovered to produce a beneficial effect in hypertensive patients as well as in hypertensive dogs. Its hypotensive action may be explained by the depression of central vasomotor reflex mechanism and adrenolytic action. Cardiac output and ganglionic transmission are not influenced. Two crystalline substances were identified as buddleoglucoside and yejuhua lactone.

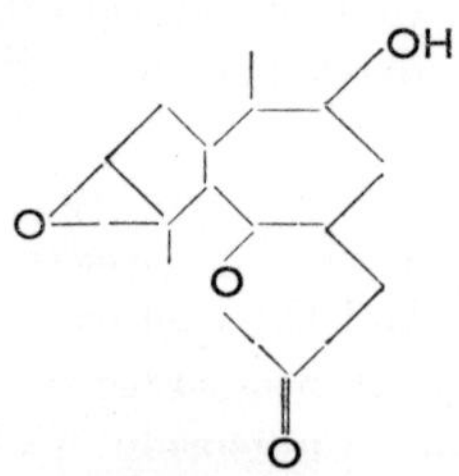

Yejuhua lactone

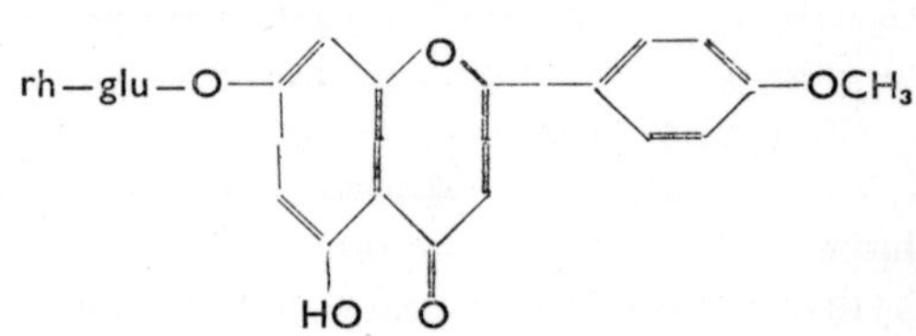

Buddleoglucoside

(e) Tienmulilmine and tienmulilminine from Tienmulilu (*Veratrum schindleri* Loes F.). The latter alkaloid is the more active antihypertensive agent. Hypotensive alkaloids have been obtained from other species of Chinese *Veratrum*. The relative potency of the total alkaloids from the three Chinese species appears in the order of *V. maackii*, *V. puberulum* and *V. schindleri*. The mode of action is mainly central vasomotor depression, as revealed by the depression of the carotic occlusion response and by the fact that vagotomy or atropine may abolish its hypotensive action. A decrease in cerebral and renal blood flow may also play a role in the hypotensive mechanism.

(f) Tu-qing-mu-xiang (*Aristolochia debilis* S. et Z.), from which debilic acid ($C_{18}H_{12}O_7N$) was recently isolated. Debilic acid appears to be closely

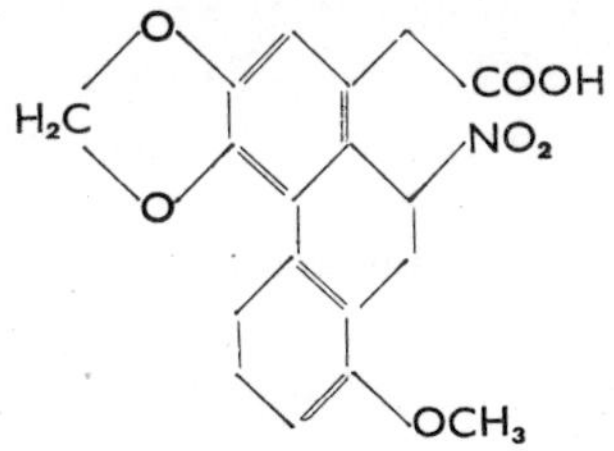

Debilic acid

related to aristolochic acid obtained by Pailer in 1956. A more active substance was later reported; its chemical properties and structure require further elucidation.

(g) Huang-qin (*Scutellaria baicalensis* Georg.) is an effective agent with low toxicity. The flavone-glucoside, baicalin, appears less active compared with a more recently obtained yellow crystalline substance.

(h) Chou-wu-tong (*Clerodendron trichotomum* Thunb.). Its hypotensive potency when assayed on hypertensive dogs is less effective than the above-mentioned *Veratrum schindleri*, *Scutellaria baicalensis* and the total alkaloids from *Rauwolfia verticillata*, but more potent than du-zhong (*Eucommia ulmoides*).

(i) Chuan-xong (*Ligusticum wallishii* Fr.) was found effective in renal hypertensive dogs and renal or cortisone hypertensive rats. It is of interest to note that this substance may potentiate the degree and duration of the hypotensive action of reserpine.

(j) Tianma-gouteng Yin is a mixture containing *Gastrodia elata* and *Uncaria rhynchophylla* as the chief constituents. Both of them exert a central depressant action, the forme being more predominant in central anticonvulsive action and the latter more as sedative–hypotensive agent. The hypotensive effect of *Uncaria* is due to both central vasomotor depression and peripheral inhibition of cardiovascular receptors, especially those of the carotid sinus. Concomitant with the fall of the blood pressure in hypertensive dogs, this mixture appears to improve markedly the disturbed activity of the higher nervous system as shown by the return of the depressed positive conditioned reflex.

Chronic hypertension is a complicated disease process. Many drugs may modify this process. By judicious combination of the various therapeutic agents, synthetic as well as naturally occurring, much progress may be expected in the future.

4. CARDIOTONICS AND DIURETICS

Investigations of cardiotonics in recent years have been centered around the study of the Chinese species of *Strophanthus* [*Strophanthus divaricatus* (Lour) Hook et Arn.] and *Thevetia* (*Thevetia neriifoia* Juss or *Thevetia peruviana* Merr.). Their active principles were compared with k-strophanthin and thevetin respectively in pharmacological analysis and clinical studies.

Divasid is the standardized preparation of the total glucosides obtained from the fruits of Yang-jiao-ao (*Strophanthus divaricatus*). The pharmacological actions of divasid are similar to those of k-strophanthin, noted for their rapid and short action. It has been used in the clinics as a substitute for k-strophanthin. Bioassay of divasid in unanesthetized pigeon with fatality as the endpoint showed a potency of 43—45 per cent of g-strophanthin (mg for mg).

Three aglucones were obtained from this species of *Strophanthus*. Strophanthiline A differs from strophanthidin by the absence of —OH radicals at C-5 and C-14 positions and the presence of —OH at C-11.

The glucosides from the Chinese yellow flower *Thevetia* are definitely different from thevetin. The potency is about four times that of thevetin, being 0·2 mg per cat unit. The following principles have been isolated: thevetin A, thevetin B, peruvoside, neriifolin, acetylneriifolin and ruvoside. Their chemical characteristics have been recently ascertained in the literature.

Other cardiotonic agents in Chinese medicinal plants investigated recently and found active include the following substances. Malijin (*Asclepias curassavica* Linn.) contains strophanthin-like substances. Lobuma (*Apocynum lancifolium* Rus.) has cymarin-like actions.

Xuban-kaikou-jian (*Tupistra fimbriata*) is a medicinal plant newly discovered; its cardioactive glucoside, tupistrin, possesses digitalis-like activity. Its potency is on the same level as the *Thevetia* glucosides.

The screening for diuretics merits some attention, since these agents are used in a variety of clinical conditions, including the late stages of *Schistosomiasis japonica* infection. Many medicinal decoctions are reputed to induce diuresis, but few have been seriously studied. The following drugs show remarkable diuretic effects in both healthy human volunteers and rabbits: *Astragalus hoantchy*, *Stephania tetranda*, *Alisma plantago*, *Daphne genkwa*, *Lobilia radicans*, Wu-ling pulvis, Shih-pi draught.

Loranthus paracticus from which a flavon-arabinoside was obtained and shown to have marked diuretic and hypotentive effects.

Phyllostachys puberla and *Lagenaria vulgaris* showed prolonged diuretic effect, especially when combined.

Atractylodes macrocephala is one of the most active diuretics out of a large number of agents screened recently in animals. An extract equivalent to 0·05—0·25 g of the crude drug per kilogram of unanesthetized dog produced a marked diuretic effect lasting more than 5 hr with 9 times the normal urine output. The mechanism of action is chiefly through the increased excretion of sodium and other electrolytes. It does not affect the creatinine clearance test and therefore its action is not related to any change of glomerular filtration process, but may be attributed to an active inhibition of the reabsorption processes in the renal tubules. It inhibits the carbonic anhydrase activity. This agent is remarkably low in toxicity.

Investigations along the line of these medicinal plants appear to be most rewarding as a lead in the search for new chemical entities in diuretics.

5. SOME NEW CHEMICAL SUBSTANCES FROM CHINESE MEDICINAL PLANTS

Cucurbitine is a new amino acid isolated from the pumpkin seeds (*Cucurbita moschata* Duch.). Its structure is beta-amino-beta-carboxyl-pyrrolidine:

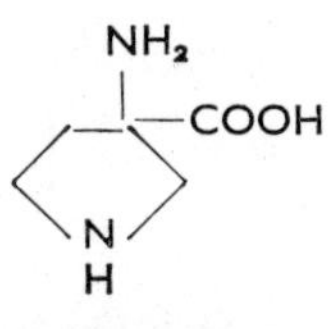

Cucurbitine

Cucurbitine is almost non-toxic, and is an effective agent in inhibiting the growth of young worms of *Schistosoma japonica*, and in producing degenerative changes in the genital organs of the worm with depressed egg-producing capacity. When combined with antimonial compound, the therapeutic effect is enhanced.

Securinine was first isolated in 1950 by Soviet scientists from *Securinega suffruticosa* (Pall) Rehd. It is a central stimulant with strychnine-like actions, but with much less toxicity. This plant is widely distributed in China. Its chemical structure has been elucidated independently and simultaneously in 1962 but along different lines of attack by Chinese and Japanese scientists.

Systematic chemical investigation of the lactones in medicinal plants showed their occurrence in a wide variety of well-known Chinese drugs.

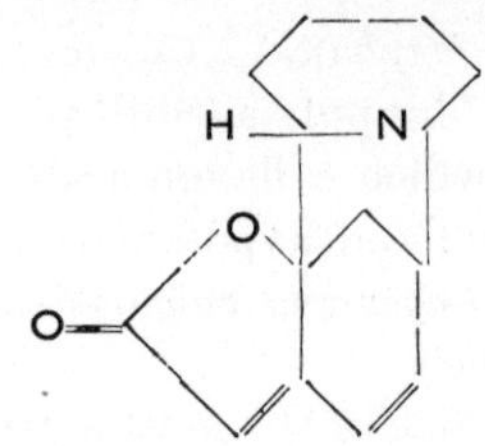

Securinine

Yejuhua lactone from *Chrysanthemum indicum*, baicalin from *Scutellaria baicalensis*, clerodendrin from *Clerodendron trichotomum*, wikstroemin from *Wikstroemia viridiflora* are some of the most recent additions to the list.

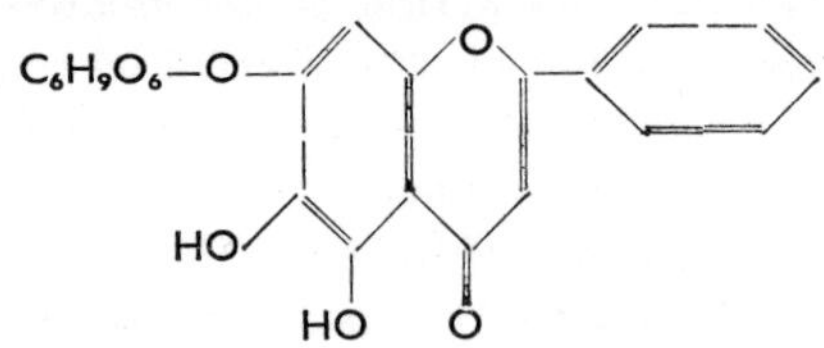

Baicalin

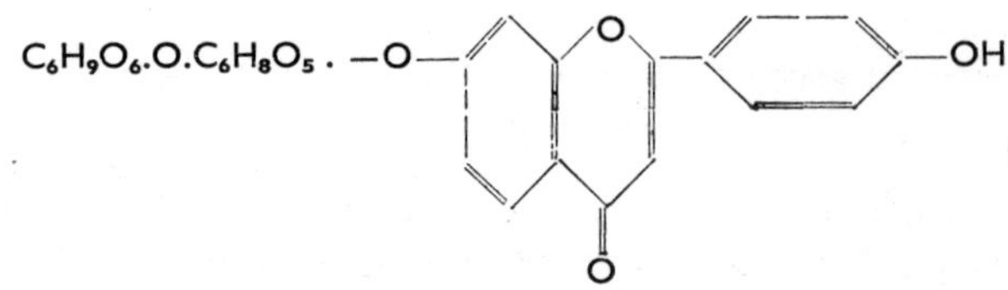

Clerodendrin

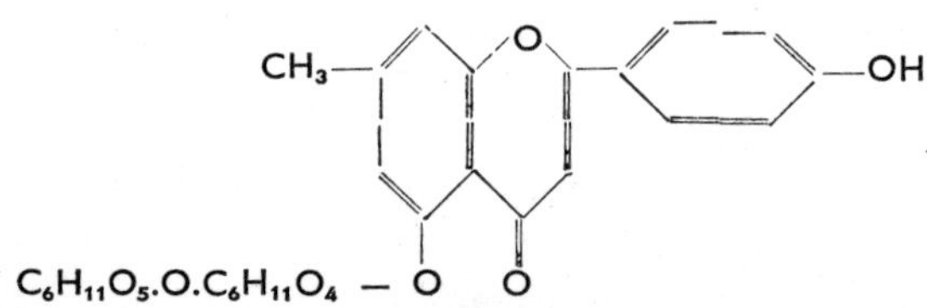

Wikstroemin

Many active substances have been obtained in pure crystalline form from Chinese medicinal plants. Some of the newest additions are:

Pseudolycorine from Chinese narcissus (*Narcissus tazatta* L. var. *chinensis* Roem.).

Hemerocallin from *Hemerocallis thunbergii* Baker.

Stachydrine from *Leonurus sibiricus* L.

Forsythin from *Forsythia suspensa* Vahl.

6. CERTAIN INVESTIGATIONS RELATED TO PHARMACEUTICAL PRODUCTION

Drug research in China is directed to socialist construction with the aim of producing more and better drugs from cheaper and domestic resources.

Much progress has been recorded along the line of research in pharmaceutical industry. In this article it may suffice to mention that the biogenic synthesis of ergometrine in artificial cultural medium has been put into production line with a ten-fold increase in the yield of the alkaloid, thus bringing to an end the history of dependence upon the importation of this drug.

Another example of outstanding success from the view-point of pharmaceutical industry is the search for botanical resources in the synthesis of the hormonal steroids. Research in steroid sapogenins was non-existent in semicolonial China. Now a number of botanical families and species, such as Dioscoreacae, Liliaceae, Agavaceae, have been found to contain profitable amounts of different sapogenins, like diosgenin, sinodiosgenin, tigogenin, hecogenin, chlorogenin, sarsasapogenin, manogenin, and a new rockogenin. Some of them are now used in the commercial synthesis of progesterone, testosterone, cortisone, prednisolone and other steroid hormones.

CONCLUSION

It is our conviction that medical science belongs to the whole of mankind, and in its development all countries have contributed their share. We too have made our contribution in the long past, and we shall have more and more creative contributions to world medicine in the future.

In the past, Chinese medicine and pharmacology were isolated and not able to profit by modern natural sciences. To explore, develop and enrich our traditional medical legacy by means of modern scientific methods can have only one outcome: advancing new frontiers of medicine heretofore unconquered by present-day medical science. The nihilistic tendency to depreciate such efforts is obviously unfounded, if not prejudicial.

This report does not include pharmacological studies of medicinal plants in Taiwan Province, an integral part of People's Republic of China, but still occupied by U. S. Imperialism. The Chinese people are determined for its liberation. We shall then be able to present a complete account of the pharmacological progress in all parts of China, Taiwan Province included.

THE MODE OF ACTION OF *SCHIZANDRA CHINENSIS*

L. VOLICER, I. JANKŮ, O. MOTL and Z. JIŘIČKA

Institute of Pharmacology and Institute of Organic Chemistry and Biochemistry, Czechoslovak Academy of Sciences, Prague, Czechoslovakia

THE FRUITS of *Schizandra chinensis* Baill. are claimed in the national medicine of the Far East to possess stimulant effects which improve physical performance and act against fatigue (Sorokhtin, 1955). In this respect it has been reported that crude extract of the fruit increased the performance of mice in the swimming test (Georgiev, 1958) and there have also been some reports of better physical performance in man (Lebedev, 1955). These experiments, however, were not made under control conditions.

As a lignan derivative, schizandrine has been recently isolated from the fruits and its chemical structure has been determined (Kochetkov, Khorlin, Chizov and Scheichenko, 1961) (Fig. 1), it seemed worthwhile to undertake some pharmacological studies using modern fractionation techniques to determine whether the central activity of the plant would be confirmed.

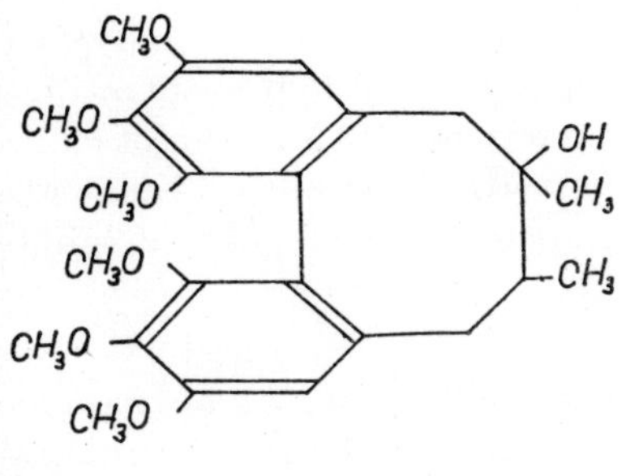

FIG. 1. The structural formula of schizandrine.

Preliminary experiments were performed with water, ethanol and petrol-ether extracts from *Schizandra* fruits. As activity was found mainly in the petrol-ether extract, only results obtained with crude and purified petrol-ether extract are reported in this communication.

MATERIALS AND METHODS

Experiments were performed on white mice of both sexes, weighing 18–22 g, and on Wistar rats of both sexes, weighing 180–250 g, in groups of at least 10 animals. As the petrol-ether extract from *Schizandra* fruits is not soluble in water, it was diluted in olive oil, mainly 1 : 10. In all quoted experiments the *Schizandra* extract has been administered orally by tube. Control groups received the same amount of olive oil.

The doses of crude *Schizandra* extract are expressed in per cent of LD_{50}, the doses of purified extract in weight units per kg.

The following drugs were used in aqueous or commercial solutions: reserpine Spofa, chlorpromazine Egyt, methylphenidate (Ritaline Ciba), nicotine bitartrate, strychnine nitrate, metrazol (Pentazol Spofa), carbachol (Enterotonin Slovakofarma) and mecholyl.

1. Acute toxicity was estimated in mice after one peroral dose of *Schizandra* extract. The lethality was observed during one week and evaluated according to Litchfield and Wilcoxon.

2. Swimming tests were performed with mice in water 29·5–30·5°C warm, according to Brekhman (1955). Mice were trained in swimming with 2 g weight on the root of the tail in single boxes 9×8·5 cm. The time of swimming for each mouse was measured and the moment when the mouse stayed 10 sec under the level of the 25 cm deep basin was taken as the end point. *Schizandra* extract was administered in doses from 7·5 to 75 per cent of LD_{50} 30 min before the beginning of swimming.

3. The lid ptosis was estimated according to Rubin and coworkers (1957). Extract of *Schizandra* was administered in the dose of 15 per cent of LD_{50} 2 hr before administration of 1 mg/kg of reserpine or 15 mg/kg of chlorpromazine i.p.

4. Catalepsy was estimated in mice by the method of Zetler and Moog (1958). *Schizandra* extract was administered in the dose of 15 per cent of LD_{50} 2 hr before administration of 1·5 mg/kg of reserpine or 3·5 mg/kg of chlorpromazine. The immobility of mouse on the vertical rod during 1 min was considered as a positive response and the percentage of positive results were calculated.

5. Duration of thiopentone anaesthesia in mice was measured after i.v. administration of 40 mg/kg of thiopentone 3 hr after 1 mg/kg of reserpine or 1 hr after 5 mg/kg of chlorpromazine i.p. *Schizandra* extract was administered 2 hr before reserpine or chlorpromazine in the dose of 15 per cent of LD_{50}.

6. Anticonvulsive action of *Schizandra* extract in mice was evaluated by the percentage of mice which were protected against the convulsive effect of 0·25 mg/kg of strychnine nitrate, 70 mg/kg of metrazole or 5 mg/kg of nicotine bitartrate. *Schizandra* extract was administered 30 min before

administration of the convulsant drugs. The results were evaluated according to Litchfield and Wilcoxon.

7. Changes in convulsive threshold for nicotine in mice were estimated by the method of continual intravenous infusion according to Hint and Richter (1958). *Schizandra* extract, methylphenidate, phenmetrazine or amphetamine were administered 30 min before the infusion of 0·05 per cent solution of nicotine bitartrate. The amount of nicotine bitartrate provoking a tonic extension of the hind limbs was measured. The antagonism against nicotine convulsion was evaluated by the percentage of mice which had their thresholds for nicotine convulsions more than twice higher in comparison with the mean of nicotine threshold in control group.

8. Antidiuretic affect of nicotine was estimated in rats by the method of Burn (1952). *Schizandra* extract was administered in the dose of 30 per cent of LD_{50} 1 hr before administration of 6 mg/kg of nicotine bitartrate s.c. Simultaneously with nicotine 5 ml/100 g of water was administered and the diuresis was estimated during every 15 min.

9. Shedding of tears was evaluated in rats after administration of 10 mg/kg of mecholyl i.p. Extract of *Schizandra* (30 per cent LD_{50}) was administered 1 hr before, and the rats were killed 5 min after administration of mecholyl. The histological sections were stained with haematoxylin–eosin and with Masson's Trichrome (modification with Fast Green FC).

10. Intestinal motility was evaluated by measurements of intestinal passage of carbo animalis in mice. Extract of *Schizandra* in the doses of 250 and 750 mg/kg was administered 30 min before administration of 15 μg/kg of carbachol s.c. Twenty minutes after administration of carbachol 0·4 ml of carbo animalis suspension in gummi arabici was administered. After next 5 min mice were killed and the length of their intestine through which the suspension passed were measured. The results are expressed in percentage of total intestinal length.

RESULTS

The acute toxicity of *Schizandra* extract is low. The LD_{50} of crude extract is 10·5 g/kg, of purified one is 2·8 g/kg. Among toxic symptoms tonic convulsions and mydriasis were observed.

In swimming performance tests doses up to 15 per cent of the LD_{50} of *Schizandra* extract seemed to increase the performance, whereas with larger doses the swimming time shortened. The variability between mice and single experiments was so large, however, that the changes are not statistically significant.

Pretreatment with *Schizandra* extract did not antagonize but significantly enhanced some central effects of reserpine, e.g. ptosis (Fig. 2), catalepsy

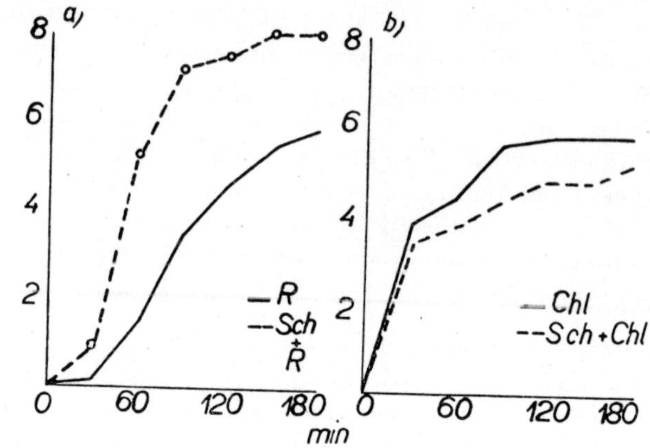

FIG. 2. The effect of *Schizandra* extract on lid ptosis in mice. (a) after 1 mg/kg of reserpine (R), (b) after 15 mg/kg of chlorpromazine (Chl). Sch, *Schizandra* extract 15 per cent of LD_{50}. Circles indicate significant changes.

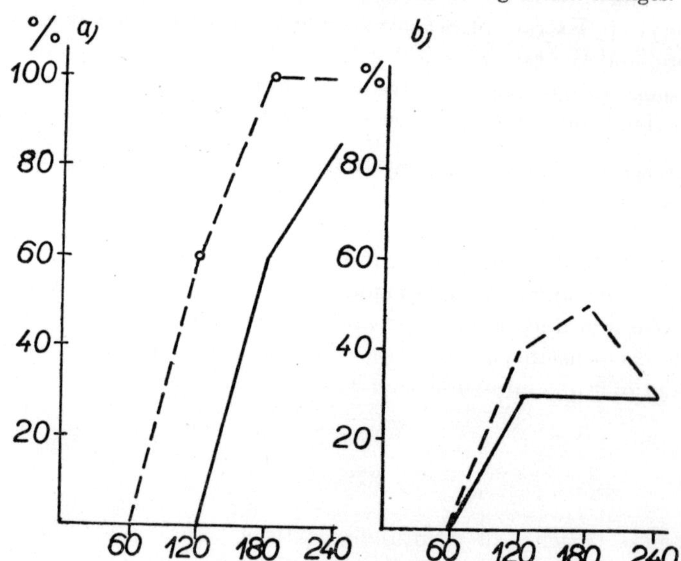

FIG. 3. The effect of *Schizandra* extract (– – –) on catalepsy in mice. (a) after 1·5 mg/kg of reserpine (——), (b) after 3·5 mg/kg of chlorpromazine (——). Circles indicate significant changes in percentage of cataleptic mice.

(Fig. 3) and prolongation of thiopentone anaesthesia (Fig. 4). On the other hand the same effects of chlorpromazine were not significantly changed.

The small doses of *Schizandra* extract lowered the threshold for nicotine convulsions (Fig. 5), whereas the larger doses had an opposite effect (Fig. 6). Methylphenidate had a similar dual dose-dependent action, while phenmetrazine only slightly raised the threshold and amphetamine was without

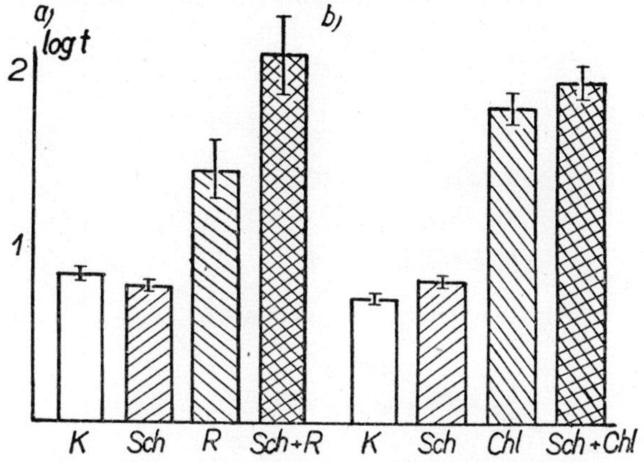

FIG. 4. The effect of *Schizandra* extract on prolongation of thiopentone anaesthesia. (a) after administration of 1 mg/kg of reserpine (R), (b) after administration of 5 mg/kg of chlorpromazine (Chl). K, control groups; Sch, *Schizandra* extract 2 hr before reserpine or chlorpromazine.

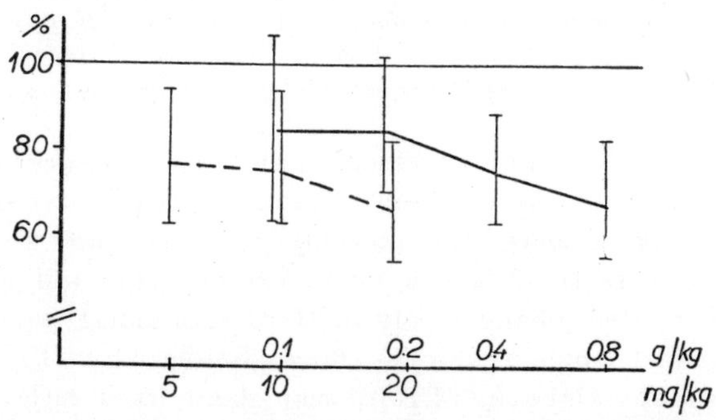

FIG. 5. The nicotine-threshold lowering effect of *Schizandra* extract and methylphenidate. The changes are expressed in percents of control group threshold. —— *Schizandra* extract; - - - methylphenidate.

effect. Anticonvulsive properties of *Schizandra* extract were most expressed against convulsions provoked by nicotine. The corresponding PD_{50} was only 25 per cent of LD_{50}. In contrast to that the PD_{50} for metrazol convulsions was in the range of the LD_{50} of *Schizandra* extract and at the same time we were not able to demonstrate a clear antagonism against strychnine convulsions.

Extract of *Schizandra chinensis* potentiated also another central effect of nicotine. After water load *Schizandra* extract itself enhanced the diuresis in rats but potentiated the antidiuretic effect of nicotine (Fig. 7). It is interesting to remark that in ethanol anaesthesia the pretreatment of rats with *Schizandra* extract did not influence the antidiuretic effect of nicotine.

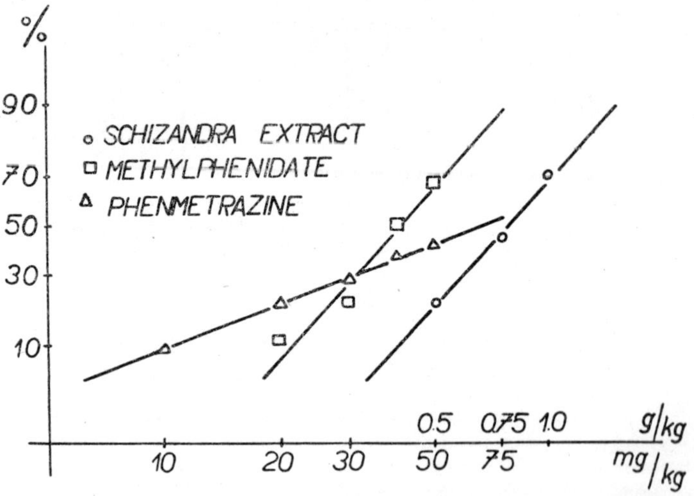

FIG. 6. The antagonizing effect of *Schizandra* extract, methylphenidate and phenmetrazine against nicotine convulsions. The percentage of mice which had their threshold more than twice higher than mean of control group were calculated according Litchfield and Wilcoxon.

Also some effects of peripheral cholinomimetic substances are potentiated by the pretreatment with *Schizandra* extract. So *Schizandra* extract increased the shedding of tears after mecholyl and the tears were bloody. In the smears of tears there was a lot of erythrocytes and leukocytes. Histologically there were changes only in Harderian gland, whereas other tear-glands remained unchanged. The microphotography (Fig. 8) shows slightly dilated tubules and acini of Harderian gland, lined with the cubical or slightly isoprismatic epithelium. Under the basement membrane we may see the fibrillar connective tissue with a small amount of either pigmented cells or free pigment granules. The secretion in the tubules and acini is rather mucous. After the application of *Schizandra* extract and mecholyl, the acini and the ducts are compressed by hyperaemia of small vessels which are in some places ruptured causing haemorhagias in the connective tissue. The lining epithelium of the acini is greatly swollen, partly disconnected and in the lumina of acini are seen free blood extravasates. Free granula of the pigment are frequently admixed to the blood. Extract of *Schizandra* alone does not influence the histological picture of Harderian gland.

The dual dose-dependent action of *Schizandra* extract may be observed on intestinal motility in the experiments with carbachol (Fig. 9). The smaller dose, which did not influence the normal intestinal motility in mice, enhanced the effect of carbachol very much, whereas the larger dose of *Schizandra* extract decreased the intestinal motility to a great extent and prevented the effect of carbachol.

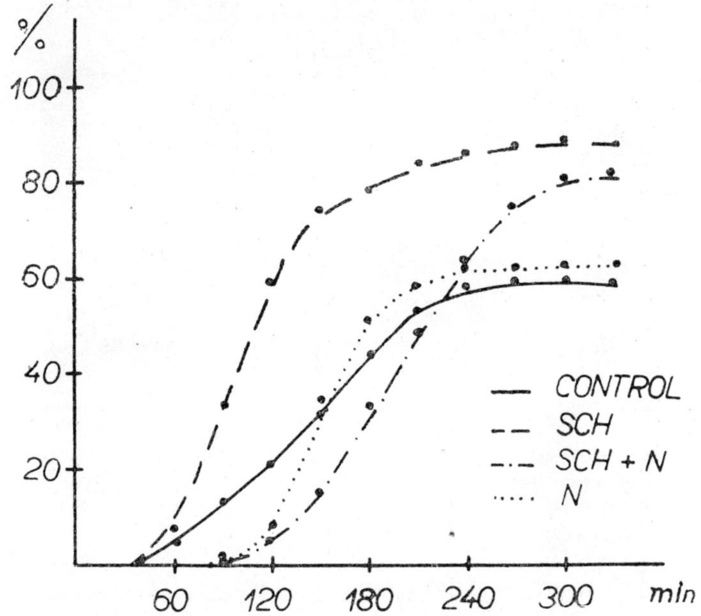

FIG. 7. The effect of *Schizandra* extract (Sch) and nicotine (N) on the diuresis after water load in rats. The diuresis is expressed in cumulative curves of urine volume in per cents of administered water.

DISCUSSION

Our results demonstrate that the petrol-ether extract of *Schizandra chinensis* in contrast to other psychomimetic drugs does not antagonize, but on the contrary potentiates the central effects of reserpine. It is not a non-specific depressive effect as the same effects of chlorpromazine are unaffected. Malone and Roth (1962) have described recently that pretreatment with yohimbine potentiates the lid ptosis after reserpine. From this point of view it is interesting to note that between schizandrine and reserpine and also between yohimbine and reserpine there are some structural resemblances.

On the other hand there is remarkable analogy of *Schizandra* extract with methylphenidate in the interaction with nicotine. A very interesting dual dose-dependent effect of both of these substances is demonstrated.

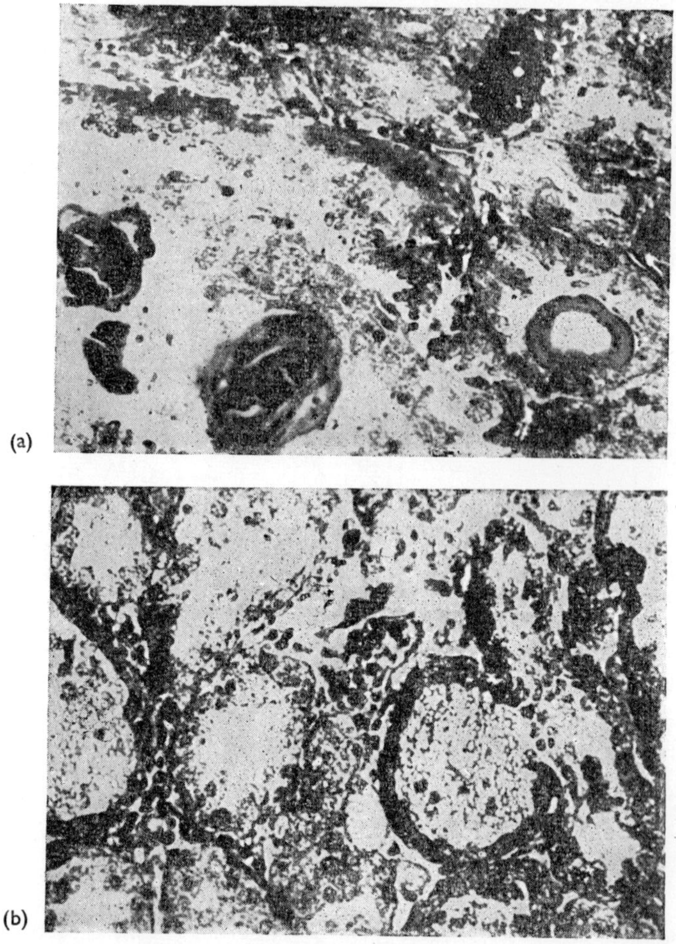

FIG. 8. Microphotograph of Harderian gland. (a) after administration of 10 mg/kg of mecholyl, (b) after pretreatment with 30 per cent of LD_{50} of *Schizandra* extract and administration of mecholyl.

In small doses *Schizandra* extract and methylphenidate potentiate the central effects of nicotine whereas in larger doses these effects are antagonized. This property of methylphenidate was described by Osborne and Sigg

(1960), who found that small doses of methylphenidate enhance the ciliary movements of the frog oesophagus while larger doses depress it. The pretreatment by *Schizandra* extract modifies also in the same way the effects of some peripheral cholinomimetic drugs. It is worthwhile to consider whether this nicotinomimetic effect is not in connection with the anti-fatigue effect of *Schizandra* fruits.

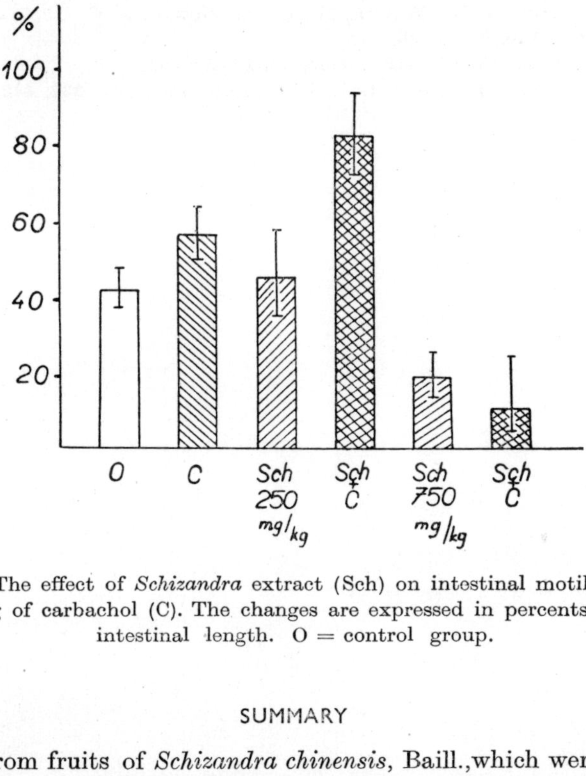

FIG. 9. The effect of *Schizandra* extract (Sch) on intestinal motility after 15 μg/kg of carbachol (C). The changes are expressed in percents of total intestinal length. O = control group.

SUMMARY

Extract from fruits of *Schizandra chinensis*, Baill.,which were used in the national medicine of the Far East against fatigue differs significantly from other psychomimetic drugs. It potentiates some central effects of reserpine, whereas the same effects of chlorpromazine remain unaffected. In small doses *Schizandra* extract potentiates also some effects of nicotine and cholinomimetic drugs, whereas larger doses the same effects antagonize. In this dual dose-dependent action *Schizandra* extract resembles methylphenidate, while phenmetrazine and amphetamine act differently.

REFERENCES

BREKHMAN, I. I. (1955) *Mat. k izuch. zhen-shena i limon.* **2**, 124—146.

BURN, J. H., FINNEY, D. J. and GOODWIN, L. G. (1952) *Biological Standardization,*

187–193. Oxford University Press, London.

GEORGIEV, V. (1958) *C. R. Acad. Bulgare Sci.* **11**, 339–342.

HINT, H. C. and RICHTER, A. W. (1958) *Acta Pharmacol. Toxicol.* **14**, 153–157.

KOCHETKOV, N. K., KHORLIN, A., CHIZOV, O. S. and SCHEICHENKO, V. I. (1961) *Tetrahedron Letters*, No. **20**, 730–734.

LEBEDEV, A. A. (1955) *Mat. k izuch. zhen-shena i limon.* **2**, 178–188.

MALONE, M. H. and ROTH, R. H., JR. (1962) *J. Pharmac. Sci.* **51**, 345–348.

OSBORNE, M., and SIGG, E. B. (1960) *Arch. Int. Pharmacodyn.* **129**, 273–289.

RUBIN, B., MALONE, M. H., WAUGH, H. M. and BURKE, J. C. (1957) *J. Pharmacol. Exper. Ther.* **120**, 125–136.

SOROKHTIN, G. N. (1955) *Farmakol. Toksikol.* **18**, 53–55.

ZETLER, G. and MOOG, E. (1958) *Arch. Exper. Path. Pharmak.* **232**, 442–458.

THE EFFECT OF AN EXTRACT OF *GINSENG* ON THE ADRENAL CORTEX

V. PETKOV and D. STANEVA-STOICHEVA

Department of Pharmacology, Postgraduate Medical Institute, Sofia, Bulgaria

FOR OUR experiments we used an aqueous-alcoholic extract (approximately 20 per cent) concentrated by evaporation of a dilute extract of the roots of cultivated Chinese *Ginseng*. In order to remove the alcohol, the extract was evaporated at a temperature of 35–40°C in a vacuum or desiccator until its volume had been halved. Subsequently the extract was diluted to its original volume with distilled water. The *Ginseng* extract obtained in this way was then given to white rats by means of a stomach tube in a dose of 0·2 ml per 100 g of body weight. Control animals received the same volume of water.

Seven series of experiments were performed on 136 rats.

1. The first experiment was designed to discover the effect of *Ginseng* on the functions of the adrenal cortex of 30 normal rats (15 of these were controls). The following criteria were used: (1) the number of eosinophil cells in the peripheral blood; (2) the weight of the adrenals; (3) the ascorbic-acid content of the adrenals; (4) the cholesterol content of the adrenals.

After determination of the initial eosinophil level by Thorn's method, 15 animals received *Ginseng* in the above-mentioned dosage for 17 days, and another group of 15 animals was given the same volume of water. On the 17th day the eosinophils were again counted and the animals were killed by decapitation. The adrenals were weighed and their cholesterol content determined histochemically by Schultz's method. The ascorbic-acid content was determined by the Giroux–Leblond method in the adrenal cortex of 10 experimental and 10 control animals. In the remaining 5 animals in each of the groups the ascorbic-acid level was chemically determined by Tilman's method.

At the end of the experiment the eosinophil count in the peripheral blood of the rats which had received *Ginseng* was 26 per cent lower than at the beginning of the experiment (the average fall was from 682 ± 38 to 504 ± 25). This fall was statistically significant (with $p = 0{\cdot}001$). In the control animals a certain reduction in the number of eosinophil cells was also observed (by 13 per cent), but this was not statistically significant (Fig. 1a).

The average weight of the adrenals of experimental animals which had received *Ginseng* was 13 per cent higher than that of control animals (per 100 g of weight) (Fig. 1b). This difference is statistically insignificant.

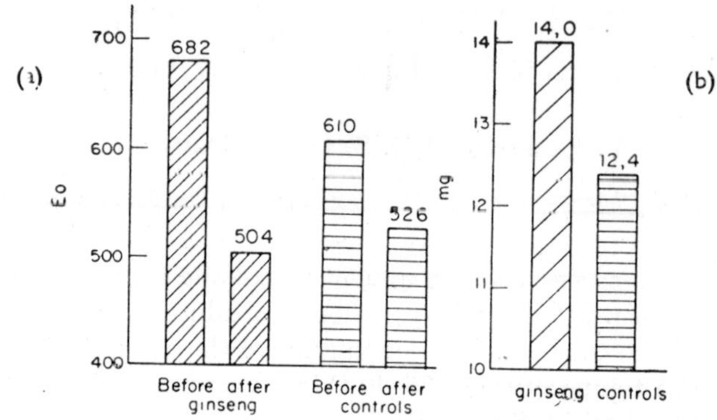

FIG. 1. Influence of *Ginseng* extract on the number of eosinophils (a) and the weight of the adrenals (b) (peroral application in the course of 17 days).

(a)

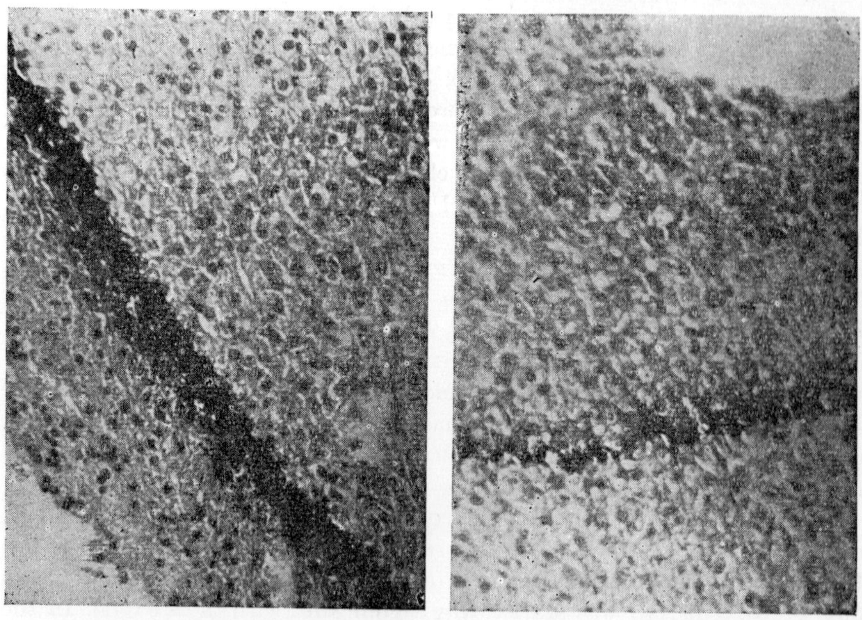

FIG. 2. Cortex of an adrenal gland stained by Girou and Leblon's method for ascorbic acid: (a) control animals; (b) experimental animals (receiving *Ginseng*).

Chemical determination of ascorbic acid did not show any substantial difference in the amount present in experimental and control animals. However, histochemical investigation showed that there was a slight de-

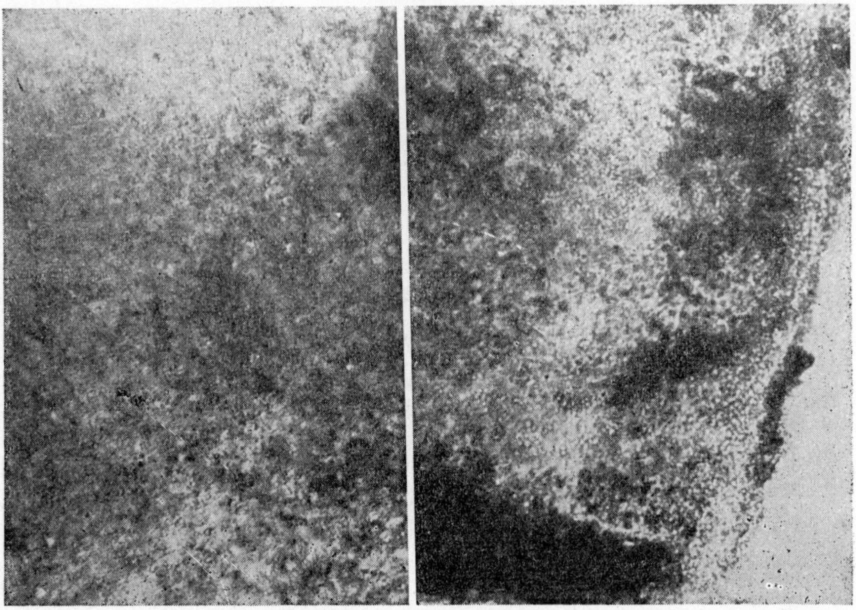

(b)

FIG. 3. Cortex of an adrenal gland stained by Schultze's method for cholesterin: (a) control animals; (b) experimental animals (receiving *Ginseng*).

crease in the ascorbic-acid content of the adrenal cortex of experimental rats as compared with that of control rats (Figs. 2a and b). The intensity of the histochemical test for cholesterol was also reduced in experimental animals (Figs. 3a and b).

2. The corticosteroid content of the urine was determined in 6 white rats by means of the Kingsley and Getchell method. Each of the rats was tested 10 times, i.e. every second day during a period of 20 days. During the following 20 days the rats received *Ginseng* extract *per os* in a dose of 0·2 ml per 100 g of body weight. The corticosteroid content of the urine was tested throughout this period using the same technique. It was found that *Ginseng* raises the amount of corticosteroid in the urine by more than 60 per cent (on an average from $0{\cdot}847 \pm 0{\cdot}053$ mg/l. to $1{\cdot}362 \pm 0{\cdot}103$ mg/l.) (Fig. 4). This increase is statistically significant ($p = 0{\cdot}001$).

3. A third series of experiments was carried out on animals under stress. In one experiment 10 experimental rats were given the above-mentioned

daily dose of *Ginseng* and 10 control rats received the corresponding volume of water. On the 11th day an eosinophil count was carried out on the animals' blood. Subsequently the animals were subjected to stress by immersing one

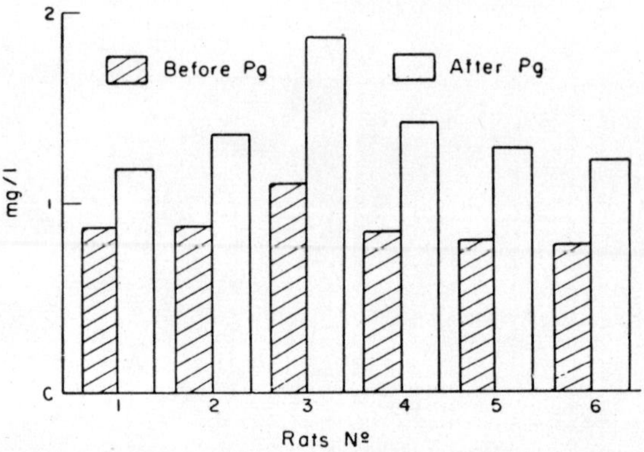

FIG. 4. Influence of *Ginseng* extract (Pg) on the amount of corticosteroids (by Kingsley and Getchell's method) in the urine of rats.

hind leg in hot water (70°C) for 1 min. The number of eosinophil cells was again determined 2 and 4 hr after this stress. The experimental animals showed a 12 per cent fall in their eosinophil count 2 hr after the stress and a 9 per cent fall after 4 hr (statistically insignificant). After the same intervals the eosinophil count of the control animals had risen by 41 and 45 per cent respectively (statistically significant) (see Fig. 5). During subsequent days 3 control rats died while all the experimental rats survived the severe stress.

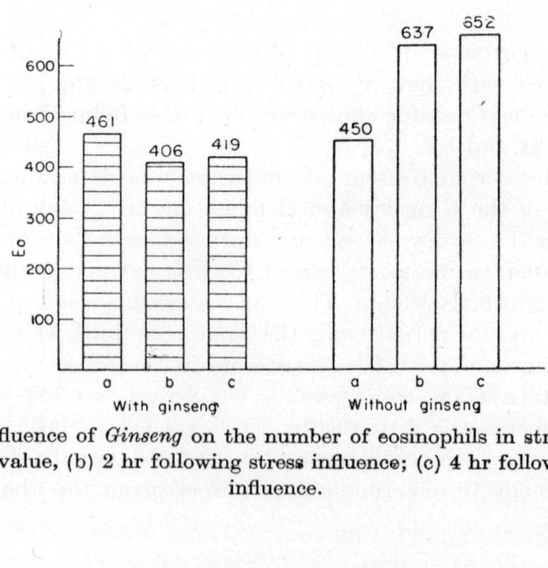

FIG. 5. Influence of *Ginseng* on the number of eosinophils in strong stress: (a) initial value, (b) 2 hr following stress influence; (c) 4 hr following stress influence.

The experiment was repeated with the same number of animals but with the following change of procedure; the experimental animals were given *Ginseng* and the control animals water on 5 instead of 10 days. On the 6th

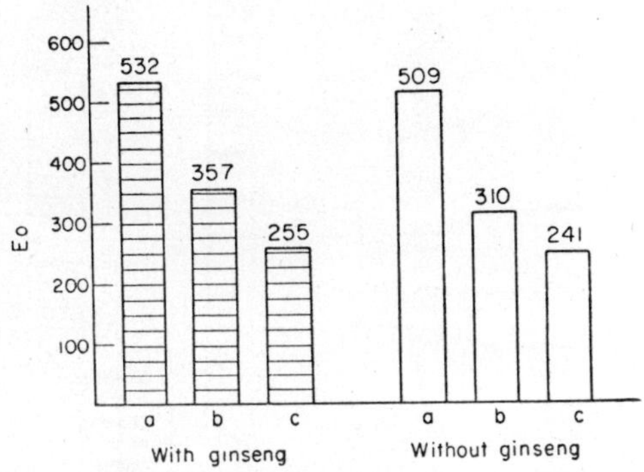

FIG. 6. Influence of *Ginseng* on the number of eosinophils in moderate stress: (a) initial value; (b) 2 hr following stress influence; (c) 4 hr following stress influence.

day they were subjected to milder stress by immersing one hind limb in hot water at 70°C for only 10 sec. Under these conditions no differences between experimental and control animals were observed. Both groups showed a decline in their eosinophil count. In the experimental group this amounted to 33 per cent after 2 hr and to 56 per cent after 4 hr. In the control group the eosinophil count had fallen by 39 and 59 per cent. This fall is statistically significant (Fig. 6).

4. In the fourth series of experiments all the rats had one adrenal removed and were divided into three groups of 20 animals each (10 experimental and 10 controls). The following criteria were used to evaluate the effect of *Ginseng* on the adrenals: (1) the degree of compensatory hypertrophy of the remaining adrenal; (2) the ascorbic-acid content of the adrenal; (3) the cholesterol level of the adrenal. After adrenalectomy the weight of the excised adrenal was determined. The other adrenal was weighed in the same way at the end of the experiment. The first group of 20 animals received the above-mentioned dose of *Ginseng* or water for 3 days, the second group for 6 days, and the third for 5 days.

The adrenal of rats in the first experimental group weighed 25 per cent more (per 100 g of live weight) than the adrenal which had been removed at the beginning. In the control group this weight increase was only 16 per cent. The difference was statistically insignificant. In the second group the intact adrenal of experimental animals weighed 46 per cent more than the adrenal which had been removed at the beginning of the experiment.

In the control group the increase was 20·5 per cent (Fig. 7). This difference is statistically significant ($p = 0{\cdot}05$). In the last group of experimental animals which had received *Ginseng* for 5 days the average weight of the

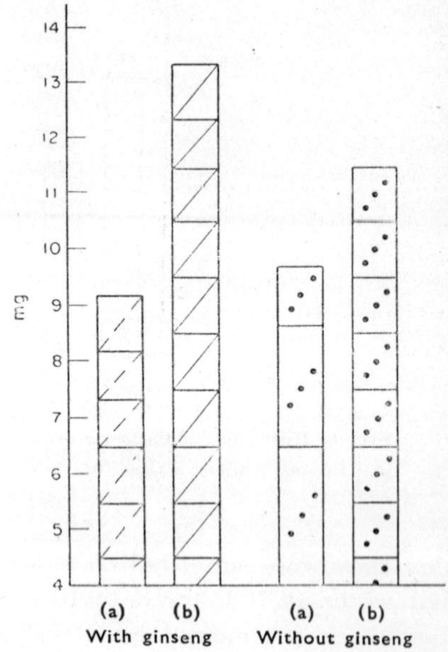

FIG. 7. Influence of *Ginseng* on compensatory hypertrophy of the intact adrenal in unilaterally adrenalectomized rats: (a) ectomized adrenal; (b) intact adrenal.

adrenals which had been left intact was 52·0 per cent higher than that of the adrenal which had been removed, whereas in the controls the corresponding increase was 40·6 per cent. This difference is statistically not significant. In this series of experiments the ascorbic acid and cholesterol content of the adrenals was determined in 5 experimental and 5 control rats. In most of the experimental animals the histochemical reaction for ascorbic acid was weaker than in the controls. Both adrenalectomized experimental and control rats had a weaker ascorbic-acid reaction than intact rats. The histochemical test for cholesterol in the intact adrenals of the experimental animals was also weaker than in the control animals.

DISCUSSION

All the data obtained indicate that the *Ginseng* extract investigated by us had a stimulating effect on the adrenal cortex. This was especially evident from the results of direct quantitative determinations of the corticosteroids excreted in the urine. These data indicate that the adaptive reaction of the organism can be actively regulated by pharmacological methods. In the experiments in which the animals were subjected to severe stress, we obtained a statistically significant increase of the eosinophil count in the control animals. This reaction was probably due to severe, rapidly appearing insufficiency of the adrenal cortex. In this case one of the most important defence reactions can lose its biological purposefulness and can become dangerous, as was found in our experiment in which 3 of 10 animals died. The experimental animals reacted quite differently to the same stress. Their eosinophil count did not rise but even fell to the same extent. It must be assumed that in this case there was no insufficiency of the adrenals in spite of the severe stress.

In view of the results obtained by one of us (V. Petkov) which have shown that *Ginseng* increases the reactivity of the cerebro-cortical cells (as can be seen from improved cortical functions, such as excitation, inhibition and mobility), we consider it very likely that the effect of *Ginseng* on the adrenal cortex is of neurogenic origin. This makes it appear that *Ginseng* may facilitate the adaptation of adrenal cortical function to the needs of the organism under changing conditions.

The our discovery of a stimulating action of *Ginseng* on glycocorticoid production throws light on its anti-inflammatory and anti-exudative effects which have been observed by other investigators.

PHARMACOLOGY OF CHINESE MATERIA MEDICA

K. K. CHEN

Indiana University School of Medicine, Indianapolis, Indiana, U.S.A.

THE CHINESE people have a history of 5000 years and their medical records extend as far back as the beginning of their history. *Pen Tsao*, the Chinese dispensatory, was written by Emperor Shen Nung, and was revised at least a dozen times when Li Shih-Cheng completed *Pen Tsao Kang Mu*. This monumental work, first published in 1956, described 1871 drugs of vegetable and animal origin. It showed that the Chinese had long discovered the cathartic properties of rhubarb, caster beans and croton oil. They had been aware of the toxicity of arsenic and mercury. Chaulmoogra oil had been considered the best remedy for leprosy. Camphor, cinnamon, and anise have been admitted to European pharmacopoeias. As medical sciences advance, investigations must be carried out with the numerous medicinal products in order to substantiate the claims of their values.

Exactly forty years ago Dr Carl F. Schmidt and I were teaching pharmacology at Peking Union Medical College, Peking, China, and began our studies on Ma Huang. We purchased the samples from a drug shop. This herb, identified as *Ephedra sinica*,[1] grows wild on the northern and western frontiers of China, and has been known in Chinese medicine since antiquity. Schmidt and I reduced the stems to a powder, made a decoction and injected a small volume into the femoral vein of an anesthetized dog. Promptly we observed a sustained rise of blood pressure. This gave us the hint that this plant contained a vasoconstricting substance. By the old fashioned method I succeeded in isolating an alkaloid by immiscible solvents. When the crystalline salt of the alkaloid was injected into another dog, the picture of our first experiment was duplicated. It was at that point that we searched the literature in our small library and found the alkaloid to be identical with ephedrine which was isolated by the Japanese Nagai[2] in 1887 from the same medicinal plant, Ma Huang. Schmidt and I did not stop there but went ahead for a thorough pharmacological and clinical investigation. We were very conscious that we were dealing with a substance similar to adrenaline, producing effects simulating sympathetic stimulation.[3] The mechanism of action is not entirely identical with that of adrenaline as well reviewed by Goodman and Gilman.[4] For example, ephedrine is not oxidized by amine oxidase but instead it exerts a competitive inhibition with the enzyme.[5] It also differs from adrenaline in that it is effective by mouth and has a more prolonged duration of action. In clinical medicine ephedrine by mouth is capable of relieving attacks of bronchial asthma.[6] The most important application of ephedrine is in spinal anesthesia. It sustains blood pressure for a long time and is a life-saving measure.[7] There have been significant developments once the clinical values have been appreciated. Originally ephedrine was manufactured from the plant shipped from China, but Hildebrandt and Klavehn[8] perfected a fermentation process which results in an optically active intermediate. The final product can therefore be obtained in optically active form without chemical resolution. The interesting feature is that ephedrine was known in pure form in 1887, adrenaline in 1901; but the latter came to medical use almost a quarter of a century before the former. At present there is a long list of synthetic derivatives of ephedrine sold on the drug market. Among them, benzedrine, dexedrine and pervitin have a potent central stimulating action. They have been used for alertness and wakefulness by pilots, motorists and others whose occupation requires long hours of physical exertion. They also seem to excite and impart thrills to their users, resulting in habituation. Thousands of people have fallen victims to this form of selfmedication, but fortunately recovered without harm because the drug is not truly addictive.[9]

Pen Tsao Kang Ma of China did not mention any species of *Digitalis*. It can be presumed that *Digitalis* is not indigenous to China, but there are products having a cardiotonic action. Chia Tso Tao, or *Nerium odorum*, is often cultivated for ornamental purposes because its flowers are both beautiful and fragrant. The recent chemical work indicates the high complexity of the glycosides.[10] We have confirmed the digitalis-like action of 16-anhydro-digitalinum verum monoacetate, odoroside D, K, odorobioside K, monoacetate of odorotrioside G, odorobioside G, and odoroside H.[11] The scarcity of these interesting compounds has prevented clinical trials.

There is a special toad which inhabits China, namely, *Bufo gargarizans*. The Chinese druggists make a commercial preparation of its parotoid secretion and the practitioners advocate its external application for the treatment of canker sores, sinusitis, toothache, and local inflammation. We isolated two cardiotonic steroids[12] for pharmacological stud es. By paper chromatography Ruckstuhl and Meyer[13] identified more than ten active substances. We made a clinical investigation with cinobufagin.[14] By intravenous injection cinobufagin can indeed reduce the ventricular rate of auricular fibrillators. In a series of six cases, a marked slow ng of pulse rate occurred after cinobufagin lasting for 5–6 hr. These patients had nausea and vomited. Therefore, a conclusion is reached that the clinical use of cinobufagin is unjustifiable for obvious reasons — lack of persistent action, occurrence of untoward symptoms, and ineffectiveness by mouth.

For a number of years, Dr T. Q. Chou of Shanghai and I collaborated with each other in our research projects. For example, his highly toxic alkaloids, peimine and peiminine of Pei Mu, have a depressant action upon the mammalian smooth muscles and the frog's heart.[15] Two alkaloids of the Chinese *Corydalis* produce catalepsy in mice and monkeys while two others cause convulsions.[16] Menisine and menisidine of Mu Fang Chi have a depressor action in cats and cause clonic convulsions in rodents and emesis in pigeons.[17]

Another Chinese herb, named Chang shan or botanically *Dichroa febrifuga*, played an important part a few years ago. The root of this plant has been known to have an antimalarial action for centuries, but never seriously confirmed by modern methods. During the last war, China ran out of the supply of quinine. The Chinese Government and military forces had to be moved to the Southwest. Malaria was widely spreading in this area. In desperation the Chinese military and governmental personnel started to use this root to control malarial attacks. Meanwhile chemical work was undertaken in different laboratories. Dr T. Q. Chou succeeded in isolating three alkaloids[18] of which the most active one is γ-dichroine. Chou supplied me with a sample of γ-dichroine which was subjected to extensive studies in laboratory animals. For example, the duckling was infected with the malarial parasite, *Plasmodium lophurae*; but, after intravenous treatment with γ-dichroine, all the parasites disappeared from red blood corpuscles. Similarly canaries infected with *Plasmodium relictum* was free from parasites after the administration of γ-dichroine.[19] Clinical trials were withheld because the alkaloid caused vomiting and hydropic degeneration of the liver.

I find it most interesting to make investigations in the field of Chinese materia medica. A good principle is to seek the satisfaction of adding knowledge to science and have a prepared mind for practical application. It is therefore particularly encouraging to me to hear Doctors Zhou and Fu speaking of their achievements in the various institutes on the mainland.

ACKNOWLEDGEMENT

This work was supported by a grant of the Lilly Research Laboratories.

REFERENCES

1. Stapf, O. Ephedra sinica. *Kew Bull.*, no. 3, 133 (1927).
2. Nagai, N. Ephedrin. *Pharm. Zeit.* **32**, 700 (1887).
3. Chen, K. K. and Schmidt, C. F. The action of ephedrine, the active principle of the Chinese drug Ma Huang. *J. Pharmacol. Exper. Therap.* **24**, 339 (1924).

4. GOODMAN, L. S. and GILMAN, A. *The Pharmacological Basis of Therapeutics,* 2nd Ed., Macmillan, New York, 506 (1955).
5. BLASCHKO, H. Amine oxidase and amine metabolism. *Pharmacol. Rev.* **4,** 415 (1952).
6. MILLER, T. G. A consideration of the clinical value of ephedrine. *Amer. J. Med. Sci.* **170,** 157 (1925).
7. RUDOLF, R. D. and GRAHAM, J. D. Notes on sulfate of ephedrine. *Amer. J. Med. Sci.* **173,** 399 (1927).
8. HILDEBRANDT, G. and KLAVEHN, W. Ephedrine. Ger. Pat. 548459, April 9 (1930).
9. LEAKE, C. D. *The Amphetamines.* Charles C. Thomas, Springfield, Ill., p. 109 (1958).
10. RITTEL, W. and REICHSTEIN, T. Die Glykoside von *Nerium odorum. Helv. Chim. Acta* **37,** 1361 (1953).
11. CHEN, K. K. and HENDERSON, F. G. Pharmacology of sixty-four cardiac glycosides and aglycones. *J. Pharmacol. Exper. Therap.* **111,** 365 (1954).
12. CHEN, K. K., JENSEN, J. and CHEN, A. L. The pharmacological action of the principles isolated from Ch'an Su, the dried venom of the Chinese toad. *J. Pharmacol. Exper. Therap.* **43,** 13 (1931).
13. RUCKSTUHL, J. P. and MEYER, K. Isolierung und Aufteilung der chloroformlöslichen Bestandteile der chinesischen Krötengiftdroge Ch'an Su. *Helv. Chim. Acta* **40,** 1270 (1957).
14. CHEN, K. K. and CHEN, A. L. Studies on toad poisons. *Arch. Int. Pharmacodyn. Thér.* **47,** 287 (1934).
15. CHEN, K. K., CHEN, A. L. and CHOU, T. Q. The pharmacological action of peimine and peiminine. *J. Amer. Pharmaceut. Assoc.* **22,** 638 (1933).
16. CHEN, K. K., ANDERSON, R. C. and CHOU, T. Q. The pharmacological action of four *Corydalis* alkaloids. *Chinese J. Phys.* **11,** 7 (1937).
17. CHEN, K. K. and CHOU, T. Q. The action and toxicity of menisine and menisidine. *Chinese J. Phys.* **11,** 29 (1937).
18. CHOU, T. Q., FU, F. Y. and KAO, Y. S. Antimalarial constituents of Chinese drug Ch'ang Shan, *Dichroa febrifuga. J. Amer. Chem. Soc.* **70,** 1765 (1948).
19. HENDERSON, F. G., ROSE, C. L., HARRIS, P. N. and CHEN, K. K. γ-Dichroine, the antimalarial alkaloid of Ch'ang Shan. *J. Pharmacol. Exper. Ther.* **95,** 191 (1949).

HYPOTENSIVE AND TRANQUILLIZING ACTIVITY OF JATAMANSONE (VALERANONE) A SESQUITERPENE FROM *NARDOSTACHYS JATAMANSI D.C.*

R. B. Arora and C. K. Arora

Department of Pharmacology, All India Institute of Medical Sciences, New Delhi, India

INTRODUCTION

The purpose of this presentation is to consider a new hypotensive and tranquillizing agent, jatamansone, which is a sesquiterpene ketone isolated from the rhizomes of *Nardostachys Jatamansi*, a plant indigenous to India. Earlier the anti-epileptic activity of jatamansone and the essential oil from rhizomes, and prolonged hypotensive activity of the essential oil have been reported (Arora *et al.*, 1958a, 1958b). Although there is still lack of universal agreement amongst the clinicians that reduction in blood pressure will provide significant benefit to the patient with hypertensive vascular disease, availability of a drug that could lower the blood pressure to normal or near normal levels without producing intolerable side effects would be of outstanding clinical interest. Carefully controlled clinical investigations of jatamansone indicate it to be a promising agent in this respect. Today, we will be mainly focusing our attention on the hypotensive and tranquillizing

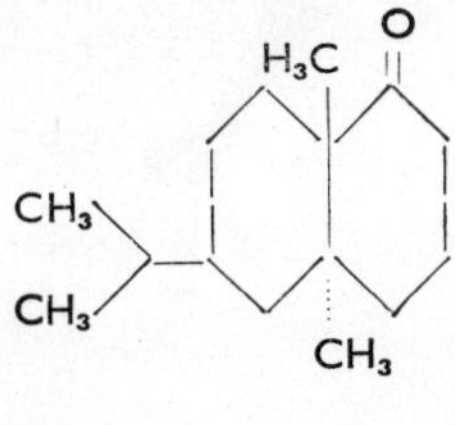

Structure of jatamansone

properties of this drug. Figure 1 shows the *Nardostachys Jatamansi* plant. It belongs to the natural order Valerianacea and is a perennial herb common in the Alpine Himalayas at altitudes of above 6000 ft. Its sable-tail, brown hairy and highly aromatic rhizomes are reported in the Ayurvedic and Unani system of medicine to be useful in the treatment of a variety of disorders such as spasmodic hysteric fits, epilepsy and other nervous disorders (Chopra, 1958) (Table 1). From the rhizomes of this herb, Gorinda-Chari (1958) isolated a sesquiterpene ketone, jatamansone.

This is synonymous with valeranone isolated by Šorm and his colleagues from Czechoslovakia.

FIG. 1. *Nardostachys Jatamansi.*

TABLE 1
USE OF N. JATAMANSI IN ANCIENT INDIAN MEDICINE

Diseases	Charak samhita 800–1000 B.C.	Sushrut samhita 800–1000 B.C.
Insanity	Chikitsa IX-45	Uttar. LX-31
Epilepsy	Chikitsa XXVI-153	Uttar. LXI-28
Skin disorders	Chikitsa VII-87	—
Snake bite	Chikitsa XXIII-54	Kalp. V-69-73
Oedema	Chikitsa XXIII-102	—
Rheumatic heart pain	Chikitsa XXIX-131	—

पुराण सर्पिर्लशुनम् हिंगु सिद्धार्थकम्वचा।
गो लोमी चाजलोमी च भूतकेशीजटा॥

S.S. Uttar. LX-46

The above potion containing *N. Jatamansi* has been used to alleviate wakefulness, overtalking, excitement and other similar mental afflictions.

Figure 2 shows the effect of the jatamansone on pentobarbital sodium hypnosis, the experimental animal being adult albino mice of either sex and the body weight between 20–24 g. The drug was given either by oral or by intraperitoneal route in doses of 1, 3, 10, 30, 100, 300 mg/kg followed 3 hr later by 35 mg/kg of pentobarbitone sodium intraperitoneally. Eight mice were used in each group.

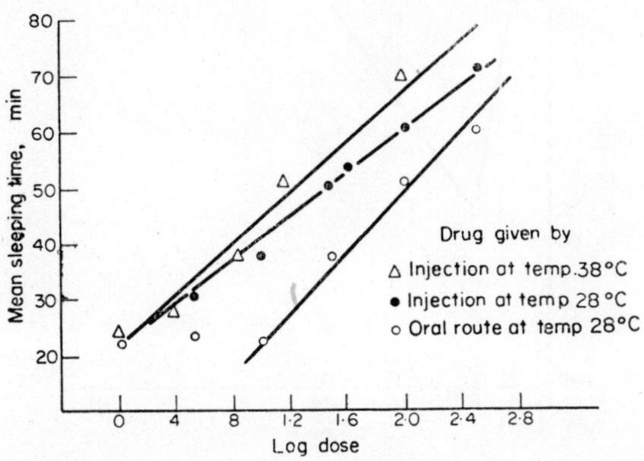

FIG. 2. The effect of jatamansone on sleeping time after pentobarbitone in mice.

One set of animals was kept at an environmental temperature of 38°C and another at 28°C in a cool room. Sleeping time (the duration of loss of righting reflex) was recorded.

Results. Graph plotted between mean sleeping time in minutes and log dose shows the effect of jatamansone. With the oral route potentiating effect was constant from the 10 mg/kg onward dose; with the intraperitoneal route the potentiating effect was constant from 1 mg/kg onward dose.

It is evident that there was no significant difference in the means between two different environmental temperatures. At environmental temperature of 38°C there is no demonstrable hypothermia but the potentiation of pentobarbital anaesthesia is significant. As suggested by Fastier *et al.* (1957), the hypnotic potentiating effect of a drug could be due to its hypothermic action. However, in this case the action of this drug is a specific one and not due to lowering of body temperature.

Figure 3 shows the effect of jatamansone on 5-HT and noradrenaline content of the rabbit's brain.

(1) For extraction of 5-HT, the method of Amin, Crawford and Gaddum (1954) was employed while assay was done by the method of Gaddum (1953). Twelve rabbits were divided into three groups consisting of four animals each. Jatamansone in a dose of 100 mg/kg was given intraperitoneally. Each group was sacrificed 4, 8 and 24 hr respectively after drug injection. Another twelve rabbits were used as control.

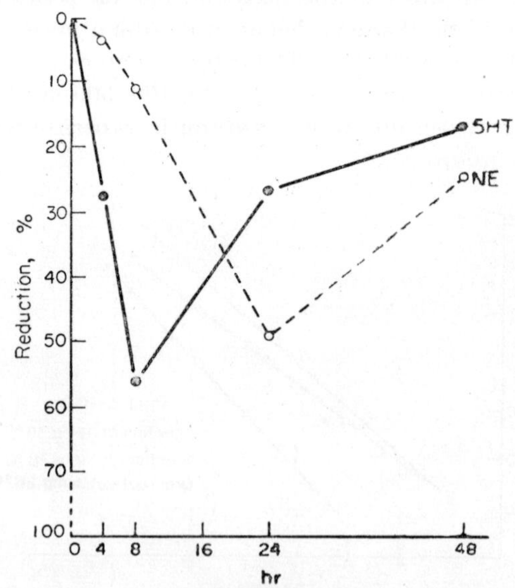

FIG. 3. The effect of jatamansone on 5-HT and noradrenaline content of the rabbit's brain.

(2) The procedure for norepinephrine extraction was essentially that of Vogt (1954), while the method of Crawford and Outschoorn (1951) was employed for assay of norepinephrine. Twelve rabbits were divided into three groups consisting of four animals each. Jatamansone in a dose of 100 mg/kg was given intraperitoneally: Each group was sacrificed 4, 8 and 24 hr respectively after drug injection. Another twelve rabbits were used as control.

Results. (1) Reduction in levels of brain serotonin started at 4 hr, reached maximum (56 per cent) at 8 hr, and is not restored to normal level up to 48 hr after drug injection.

(2) Reduction in the level of brain norepinephrine started at 4 hr and reached a maximum at 24 hr. It did not reach normal up to 48 hr after drug injection.

From the results it is evident that the time course of the effect of jatamansone on serotonin tissue levels and on norepinephrine tissue levels are distinctly different. The maximum serotonin depletion is at 8 hr after drug injection and this corresponds with the peak of tranquillization obtained by the drug as shown by conditioned avoidance response in rats. The maximum norepinephrine depletion is at 24 hr, the time at which tranquillization produced by the drug is over. The intimate relationship between tranquillization produced by a drug and its effects on brain chemistry has earlier been stressed by a number of workers. In view of the recent publication of Brodie *et al.* (1960) and Costa *et al.* (1962), it appears that a characteristic in common with reserpine-like tranquillizing agent is the production of tranquillization, only with 50 per cent or more depletion of brain serotonin irrespective of the depletion of norepinephrine. This difference in the time courses between serotonin and norepinephrine suggests that mechanisms other than inhibition of decarboxylase may be involved, particularly with respect to the effect on norepinephrine.

Our next step was to evaluate whether or not the drug has got any toxic effect. Jatamansone was given daily at a 100 mg/kg dose in rats for 5 weeks and the animals were sacrificed at the end of the period. Brain, liver, spleen and cardiac muscle were examined for signs of toxicity.

Results. Brain, cardiac muscle and spleen showed no abnormalities. On liver there was slight cloudy swelling and fatty degeneration near the centrilobular region and no evidence of damage at the peripheral regions. Haemorrhage was not seen. On kidney there were healthy glomeruli, but slight cloudy swelling was present in some areas in the urinary tubules.

At lower doses, viz. 30 mg/kg, there was no evidence of toxicity. This dose is 10–13 times more than the minimal effective dose in hypertensive patients. The toxicologic studies presented in this data have not revealed any contra-indication to the use of this drug in man. The margin of safety may be considered wide.

Jatamansone was administered to man subsequent to uneventful subacute animal toxicity studies. The first step was to evaluate the minimal effective dose and the duration of action of a single minimal effective dose in twelve hypertensive men studied as outpatients. After establishing this, hypotensive response of this agent was studied in ten hospitalized patients with hypertension (Table 2).

Patients selected were cases of hypertension of long duration who had either not shown satisfactory response to reserpine or to ganglion blockers because of troublesome side effects of the latter. Table 2 gives a summary of the clinical findings in ten hospitalized patients. In one patient (No. 1) there was evidence of left ventricular hypertrophy. One patient (No. 5) gave history of angina of effort. According to Keith Wagner's classification, two had Grade I (Nos. 5 and 6), three had Grade II (Nos. 1, 4 and 7) and two had Grade IV (Nos. 8 and 9) retinal vascular changes. Fundus was normal in the remaining 3 patients.

Lying and standing blood pressure was measured twice daily and a single average figure in each position was calculated for each day. After a control period of 7 days on placebo medication, jatamansone was given orally for a period of 1 to 6 weeks in single doses ranging from 2·5 to 10 mg per day.

The effect of jatamansone on the recumbent blood pressure is shown in Fig. 4. The top of each bar represents systolic blood pressure and the

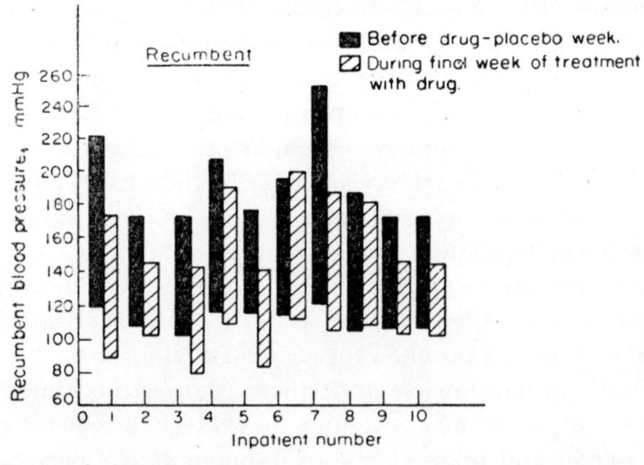

FIG. 4. The effect of jatamansone on the recumbent blood pressure in man.

bottom represents diastolic. Each solid bar represents patient's average recumbent blood pressure during the control week of placebo medication; the adjacent hatched bar represents the average pressure during the last 7 days of jatamansone administration. It is apparent that the recumbent blood pressure was lowered in nine of the ten patients.

TABLE 2. GROUP OF PATIENTS WITH HYPERTENSION

Case No.	Age, sex	Diagnosis	Average supine	Initial B.P. standing	Left ventricular hypertrophy (clinically, E.C.G., X-ray chest)	Blood urea	Albumin (urine)	Fundi	Arterial thickening
1.	40 F.	Benign essential hypertension with early left ventricular failure	221/119	211/112	Yes, present left ventricular hypertrophy	20 mg%	NIL	Grade II	NIL
2.	35 M.	Benign essential hypertension	173/110	163/101	No evidence	40 mg%	NIL	Normal	NIL
3.	50 F.	Benign essential hypertension	170/103	162/100	No evidence	18 mg%	NIL	Normal	NIL
4.	45 F.	Benign essential hypertension	209/118	199/110	No evidence	28 mg%	NIL	Grade II	NIL
5.	48 M.	Benign essential hypertension, angina pectoris	174/114	×	No evidence	25 mg%	NIL	Grade I	NIL
6.	55 M.	Benign essential hypertension	194/115	184/108	No evidence	30 mg%	NIL	Grade I	NIL
7.	38 M.	Benign essential hypertension	251/122	224/110	No evidence	26 mg%	NIL	Grade II	NIL
8.	28 M.	Malignant hypertension	177/108	165/104	No evidence	140 mg%	++	Grade IV papilloedema	Present
9.	26 M.	Malignant hypertension	165/106	155/104	No evidence	80 mg%	+	Grade IV papilloedema	Present
10.	25 M.	Benign essential hypertension	173/110	163/101	No evidence	20 mg%	NIL	Normal	NIL

Figure 5 shows the effect of jatamansone on standing blood pressure, in the same fashion as described for the recumbent pressure.

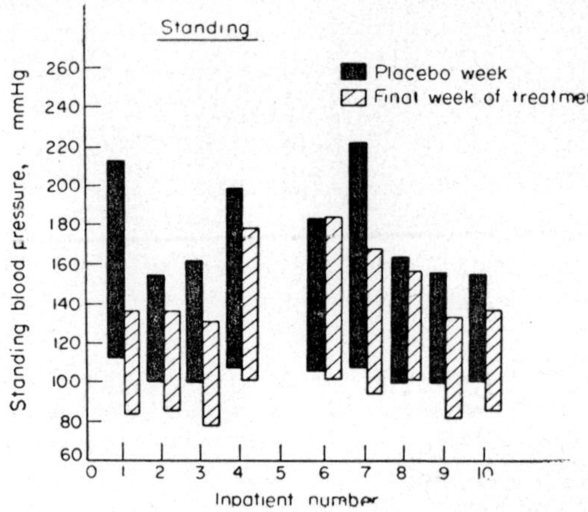

FIG. 5. The effect of jatamansone on standing blood pressure in man.

A definite fall in standing blood pressure occurred in eight out of the nine patients. In the tenth patient (No. 5), standing blood pressure was not recorded, since he was a patient of angina of effort.

Since norepinephrine is the transmitter of the sympathetic nervous system, it can be speculated that consistent reduction of standing blood

TABLE 3

PERCENTAGE CHANGE IN BLOOD PRESSURE DUE TO JATAMANSONE TREATMENT

	Values	Recumbent	Standing
Systolic B.P. mmHg	No.	10	9
	Mean	12·45 ± 3·19	15·15 ± 4·06
	"t"	3·91	3·71
	P	<0·1†	<0·01†
Diastolic B.P. mmHg	No.	10	9
	Mean	11·50 ± 3·21	13·65 ± 2·88
	"t"	3·21	4·74
	P	0·02*	<0·01†

* = Significant at less than 2% level.
† = Highly significant at less than 1% level.
± = S.E. standard error of the mean. "t" = Tabular value of *t*.
P = Probability that the differences were random fluctuations of the sample with 1 = 100 per cent. *P* of 0·05 or less was considered significant.

pressure by jatamansone may be due to a reduction in norepinephrine biosynthesis. Table 3 shows the results of statistical analysis of our clinical studies on jatamansone. Percentage change in systolic and diastolic blood pressure in recumbent and standing position is shown. Reduction in systolic pressure was highly significant in recumbent ($P < 0{\cdot}01$) as well as standing position ($P < 0{\cdot}01$). Change in diastolic pressure was significant in recumbent position; in standing position it was highly significant.

CONCLUSION

The prolonged hypotensive effect of the drug in dogs, rats and specially so in hypertensive men was found to be in several times smaller doses (5 mg) as compared to the tranquillizing dose in animals (100 mg/kg). Jatamansone significantly reduced the 5-hydroxytryptamine and norepinephrine contents of rabbit's brain and catecholamine content of rabbit's adrenal glands. The decrease in 5-HT content appears to be due to a decreased activity of 5-hydroxytryptophan-decarboxylase, while the M.A.O. activity in brain remains unaltered following jatamansone administration. Aromatic amines impinge on almost every area of biological and medical interest. These studies indicate a relationship between amine metabolism and blood pressure regulation (Table 4). The mild tranquillizing effect of the drug is

TABLE 4

THE EFFECT OF JATAMANSONE ON 5-HYDROXYTRYPTOPHAN DECARBOXYLASE, MONOAMINE OXIDASE AND 5-HYDROXYINDOLEACETIC ACID

5-hydroxytryptophan decarboxylase			Monoamine oxidase			5-hydroxyindoleacetic acid (5-HIAA)		
Vol. CO_2 evolved/g tissue/hr		% inhibition	Vol. O_2 uptake/g tissue/hr		% inhibition	mg/24 hr urine		% decrease
Normals (10)	Treated (10)		Normals (10)	Treated (10)		Normals (10)	Treated (10)	
79·7 ± 2·05	57·8 ± 2·71 $P < 0{\cdot}01$	27	150·6 ± 1·85	151·9 ± 1·89 $P >$	NIL	0·49 ± 0·06	0·17 ± 0·03 $P < 0{\cdot}01$	66

Figures in parentheses denote the number of animals in each group ± standard error of the mean.

5-hydroxytryptophan decarboxylase and monoamine oxidase assays were carried out after 8 hours of 100 mg/kg body weight jatamansone administration, while 5-hydroxyindoleacetic acid estimation was on 24 hours urine sample.

also in line with current thoughts concerning the role of amines in brain function. Intraperitoneal LD_{50} of jatamansone calculated by the method of Burn, Finney and Goodwin is 580 ± 6 mg/kg. The effect of jatamansone on blood pressure has been of sufficient magnitude to warrant extensive therapeutic application of this compound.

ACKNOWLEDGEMENTS

We are indebted to the Indian Council of Medical Research, New Delhi, for financial assistance. We are grateful to Col. M. S. Rao, Chief Physician, Safdarjang Hospital, New Delhi, for his supervision and able guidance in conducting clinical trials of this drug.

The supply of genuine samples of *Nardostachys jatamansi* by Dr. Captain of the Himalaya Drug Co. Bombay, for extraction of jatamansone, is gratefully acknowledged.

REFERENCES

AMIN, A. H., CRAWFORD, T. B. B. and GADDUM, J. H. (1954) *J. Physiol.* **126**, 596.

ARORA, R. B., SHARMA, P. L. and KAPILA, K. (1958a) *Ind. J. Med. Res.* **46**, 782.

ARORA, R. B., SINGH, K. P., DAS, P. K. and MISTRY, P. N. (1958b) *Arch. Int. Pharmacodyn.* **113**, 367.

BRODIE, B. B., FINGER, K. F., ORLANS, F. B., QUINN, G. and SULSER, F. (1960) *J. Pharmacol.* **129**, 250.

BURN, J. H. (1952) *Biological Standardization*, p. 120, Oxford Medical Publication.

CHOPRA, R. N., CHOPRA, I. C., HANDA, K. L. and KAPUR, L. D. (1958) *Indigenous Drugs of India*, 2nd ed. p. 515.

COSTA, E., GESSA, G. L., HIRSCH, C., KUNTZMAN, R. and BRODIE, B. B. (1962) *Ann. N.Y. Acad. Sci.* **96**, 118.

CRAWFORD, T. B. B., and OUTSCHOORN, A. S. (1959) *Brit. J. Pharmacol.* **125**, 353.

FASTIER, F. H., SPEDEN, R. N. and WAAL, H. (1957) *Brit. J. Pharmacol.* **12**, 251.

GADDUM, J. H. (1953) *J. Physiol.* **119**, 363.

GOVINDACHARI, T. R., RAJADURAI, S. and PAI, B. R. (1958) *Chem. Ber.* **91**, 908.

VOGT, M. (1954) *J. Physiol.* **123**, 451.

ANTI-CANCER ALKALOIDS OF *VINCA ROSEA*

R. L. NOBLE

Cancer Research Centre, University of British Columbia, Vancouver, Canada

THE PLANT *Vinca rosea* Linn. (periwinkle) of the family Apocynaceae grows as a perennial in subtropical regions throughout the world. In more northern regions it enjoys a reputation in gardens and is grown as an annual with attractive pink, red or white flowered varieties being available. Svoboda *et al.*[1] have pointed out, however, that the probable correct botanical name for this plant is *Cantharanthus roseus* G. Don. Although not strictly an oriental medicinal plant, it has held a popular reputation in folklore medicine throughout the world. Reports are available of its use in South America,[2] West Indies,[3] Africa,[4, 5] Australia,[6–8] and the Philippines.[5] Most of these reports have centred on the oral use of water extracts of the plants for the treatment of diabetic hyperglycemia and in at least two countries preparations have been marketed under such names as "Covinca" and "Vinculin".

The use of extracts of *Vinca rosea* by diabetics in Jamaica was originally brought to our attention in 1949 and led us to orient our original studies towards a search for a hypoglycemic factor in the plant.[9] The failure of our tests was communicated in the same year to Dr. C. D. Johnston of Jamaica who had supplied the plant material used by diabetics under his care. The widespread use of the plant by diabetics in many different parts of the world raises a point of caution in condemning its use from animal tests exclusively. However, there is little support from various investigators to indicate any effect of extracts on carbohydrate metabolism in animals.[5–9]

In these experiments other side effects which were curiously specific and severe came to our attention. Following a single injection of the crude material many rats died of widespread infection which was always associated with a severe depression of the white blood cells and their parent cells in the bone marrow. The consistency of these marked changes suggested using the reaction as an assay method in attempts to isolate the active substance from the plant. Serious attempts to fractionate extracts were not initiated until Dr. C.T. Beer joined the staff of the Collip Medical Research Laboratory in London, Canada, in 1954. Although at that time it was reported to the National Cancer Institute of Canada that crude extracts injected into rats caused a softening and necrosis of some types of sarcoma. Extensive studies on tumours were not undertaken until the alkaloid Vincaleukoblastine, isolated by Dr. Beer was reported in 1958.[9] At this time it was learned that a group at Eli Lilly, Indianapolis, including Drs. I. S. Johnson and G.H. Svoboda, had been working independently on the plant also investigating its reported hypoglycemic properties. They had noted that extracts, although not affecting carbohydrate metabolism, were capable of prolonging the life of mice innoculated with P1534 leukemia.[10] The active participation of a group of biologists and chemists from this laboratory, and particularly Drs. N. Neuss, M. Gorman and G. H. Svoboda, has culminated in an extremely impressive series of isolation and structural identification of many alkaloids of a new type. Of these, four closely related dimeric indole, dihydro indole derivatives have been shown to exert specific carcinostatic properties (vincaleukoblastine, vinblastine or VLB; leucocristine, vincristine or VCR; vinleurosine; and vinrosidine), and two of these have been made available in quantity for experimental clinical chemotherapeutic studies (vinblastine and vincristine — cf. Fig. 1).[5] The historical aspects have

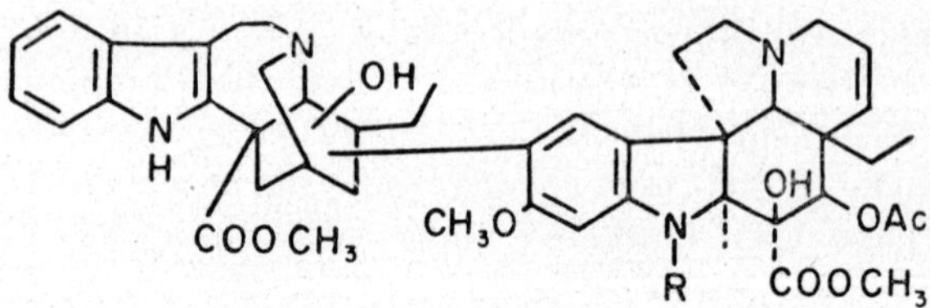

Fig. 1. The structural formulae of vinblastine ($R=CH_3$) and vincristine ($R=CHO$).

been extensively reviewed, so that the remainder of this paper will present a more general discussion but will include various unreported aspects of this problem which are in progress in our laboratory at the present time.

The somewhat unorthodox and serendipidous approach to this aspect of cancer chemotherapy has been emphasized.[9, 11] It may be of particular interest to pharmacologists to note in retrospect that the isolation of an active anti-cancer agent was accomplished through an assay procedure based on a pharmacological action on bone marrow rather than one using a tumour system. (Although this was ideal for detecting the activity of VLB it is questionable whether VCR could have been isolated by such an assay procedure since its effect on bone marrow is considerably less.) Of some importance to the cancer chemotherapist is the curiously specific action of the alkaloids against certain tumours in mice which probably would have failed to lead to their recognition had ordinary methods of tumour screening been originally employed. All recent studies have emphasized the remarkably different biological properties of compounds so closely related that they may differ only slightly. Thus VLB and VCR differ only in that the anilino-nitrogen of the dihydro indole portion of VLB carries a $-CH_3$ group and that of VCR a $-CHO$ group (Fig. 1). Yet the former compound has little effect on acute leukemia in humans, although markedly depressing the bone marrow, whereas the latter is a highly effective agent against acute leukemia in children but has little toxic side effects on the bone marrow, but uniquely produces damage to neuromuscular tissue. Some challenging inconsistencies have arisen in the experimental observations. Originally the injection of sterile crude extracts of the plant was followed by generalized infection and multiple abscesses in rats, whereas the same effects have seldom been duplicated by the purified alkaloids, although profound leukopenia may occur. In Dr. Beer's original method of separation of VLB, definite activity, presumably attributable to other unidentified alkaloids, still remained in the "strip" fraction of the column. In a number of experiments a reversible anti-folic acid type of effect was found in these, but has not subsequently been demonstrated in the purified compounds.[9] The mode of the anti-tumour action of the vinca alkaloids still remains an enigma, although it would appear unrelated to that of other types of carcinostatic agents. These and many other unanswered problems would seem to be of importance in order to assess the chemotherapeutic potential of the alkaloids. One cannot help but be overawed at present by the endless stream of new chemotherapeutic agents being made available. Some of these appear to catch the clinician's fancy only to be rapidly discarded as fresh leads appear. The impetus to find newer and better compounds may leave many potentially effective compounds incompletely investigated. Until the metabolism of new compounds is understood and some knowledge of their mode of action is known it is obviously impossible to ascribe the optimal dose levels, spacing of dosage, route of administration, etc., not to mention adequate means of preventing or reversing toxic side effects. Admittedly our knowledge of such factors in the case of the vinca alkaloids is scanty and the investigative side has been difficult. Some progress, however, has been made and observations which have been of particular interest to us may be outlined.

STUDIES ON TOXICITY

Mice. In early studies it was apparent that the mouse was considerably more tolerant to VLB than was the rat so that the former species could receive from 3–6 times the dose on a body weight basis. The tolerance of the mouse probably accounted for the better anti-tumour response in this species, since larger doses of VLB could be administered. In general it has been observed that single injections have been tolerated much better than if the doses are given repeatedly at 24-hr intervals. The MLD_{50} has varied up to 300 per cent when different strains of mice have been used. One striking feature of single injections of VLB has been the apparent sensitivity to intraperitoneal injection and relative inactivity of oral administration as may be seen in Table 1.

TABLE 1

VLB TOXICITY—FEMALE SWISS MICE

0·1 ml/10 g body wt., wts. 19—24 g

Route	Dose mg/kg	No. mice	Per cent mortality in 2 weeks
i.p.	10	10	100
	6—7·5	20	95
	4—5	20	35
	2·5	10	10
i.v.	35	5	100
	30	4	100
	20	5	20
	15	5	0
s.c.	30	5	20
	15	5	0
	7·5	5	0
oral	30	10	0
	15	10	0

It will be noted that mice will tolerate almost 10 times the amount of intravenously injected VLB when compared with the intraperitoneal route. Intraperitoneal injections are better tolerated by mice which bear well-developed ascitic tumours as seen in Table 2, suggesting some direct fixation or inactivation of the alkaloid by the tumour cells in the peritoneal cavity.

Rats. The cause of death following VLB has been most extensively studied

TABLE 2

VLB TOXICITY — NORMAL VS. 7-DAY EHRLICH ASCITES

Route	Dose mg/kg	Per cent mortality in 3 days	
		Normal	With ascites
i.p.	12.5	100 (9)	25 (8)
	7·5	50 (10)	11 (9)
i.v.	35·0	100 (5)	20 (5)

Number of mice in brackets.

in the rat. Mortality may be divided into at least two phases. Doses of the alkaloid which are just tolerated lead to marked loss of weight, diarrhoea and marked leukopenia. Some animals may succumb some 5–14 days after a single injection and it would appear that a loss of general resistance to infection is a major factor causing death. Larger doses of VLB, however, show an acute toxicity so that animals may die from 6–18 hr after injection. At this time they show an essentially normal W.B.C. and bone marrow. The thymus may involute rapidly and be haemorrhagic. Haemorrhage may also occur in the small bowel. In many cases death appeared to be related to the development of cardiac or respiratory distress with the onset of cyanosis. The heart was frequently distended and congested. Experiments by Dr. Lotz did not indicate that transfusion of autologus or isologus bone marrow or fetal spleen cells administered 24 hr after the alkaloid affected the toxic effects of intravenous VLB. Previously reported experiments had suggested some protection when VLB was administered intraperitoneally.[11] The final assessment of this type of experiment will await knowledge of the duration of action of the injected VLB which may affect the transfused cells. Earlier observations had noted some variation on toxicity between male and female rats when VLB was administered intraperitoneally[11] and Dr. Lotz's more recent studies have shown a clear-cut sex difference in the rat, but also an effect of age when the alkaloid is given into the tail vein. In immature rats both sexes showed essentially the same mortality after a single intravenous injection of varied doses of VLB. Old female rats, however, showed a surprising tolerance to the drug, so much so that little mortality followed a dose from which all young animals succumbed. This data may be seen in Table 3. Similar but less striking differences have been found in mice.

TABLE 3

MORTALITY AFTER INTRAVENOUS VLB SO_4 IN RATS

	Per cent mortality (No. rats)					
	Weight 60–80 g		Weight 140–160 g		Weight 200–240 g	
VLB SO_4 mg/kg	M	F	M	F	M	F
2·5	60 (10)	60 (10)		40 (10)	20 (10)	10 (10)
5·0	100 (10)	100 (10)	100 (15)	70 (10)	100 (10)	30 (10)
7·5				100 (20)		53 (15)
12·5						90 (10)
15·0						100 (10)

These changes suggested the possible implication of sex hormones affecting toxicity so that suitable groups of rats were treated with various hormones and the results shown in Table 4. From these observations it was apparent

TABLE 4

EFFECT OF STILBESTROL ON MORTALITY AFTER INTRAVENOUS VLB

Treatment	% mortality (No. rats)	Mean days survival
Female rats 60–140 g – VLB 5 mg/kg		
Controls – sesame oil	100 (20)	2·6
Stilbestrol × days		
20–100 γ daily × 4	93 (14)	4·1
20–100 γ daily × 7	60 (10)	8·0
20– 40 γ daily × 7	40 (10)	17·0
20– 40 γ daily × 14	0 (10)	30·0
Female rats 160–230 g – VLB 15 mg/kg		
Controls – sesame oil	95 (19)	1·5
Stilbestrol × days		
40 γ q. 2 days × 7	20 (10)	7·0
100 γ daily × 14	10 (10)	8·0
500 γ daily × 7	60 (10)	7·0
500 γ daily × 14	20 (10)	15·0

that treatment with stilbestrol would enhance the tolerance of male or female rats to VLB. Curiously it was noted that pretreatment was essential and required some 7–14 days before the protective effect was established. Some preliminary experiments have indicated a similar, but less striking, protective action of stilbestrol against the toxicity of VCR. In this case, the relatively greater potency of the alkaloid was apparent since a single intravenous dose of 2·5 mg per kg caused 100 per cent mortality in control animals. With this route of administration VCR was approximately 3–5 times more toxic than VLB. The way in which estrogens reduced the toxicity of VLB has not been explained. However, it does not appear that increased destruction of the alkaloid takes place since the expected haematological and anti-tumour responses may still be elicited in stilbestrol-treated animals. The effect of androgens on toxicity of VLB has not yet been determined. However, it has been noted that rats castrated when immature did not show an altered sensitivity to VLB. Estrogen treatment of castrate males did appear to confer greater protection than in control male animals. These observations have not been applied to the treatment of cancer in humans but Armstrong[12] has observed an apparent unexpected toxicity of VLB in a patient with breast cancer who had been receiving injections of testosterone propionate.

Some rats surviving after large doses of VLB and VCR were noted at 3–5 weeks to suddenly lose weight. In these animals the curious observation was made that the upper and lower incisor teeth had broken off at a darkened area in the tooth. It appeared that the growing end of the tooth had been affected and weakened at the time of injection and after some weeks growth broke off, making it difficult for the animal to gnaw the hard cubes of food. New normal tooth regeneration was obvious beneath the affected area. The substitution of soft food for the cubes allowed the rats to regain their weight loss. Anaesthetics, particularly barbiturates, have been noted to markedly increase the mortality of intravenously injected VLB. (Injections into the tail vein, femoral vein, femoral or carotid arteries were well tolerated and gave rise to no significant differences.)

The typical haematological response to intraperitoneal VLB in the rat may be seen in Fig. 2 and has been previously described.[9] Although

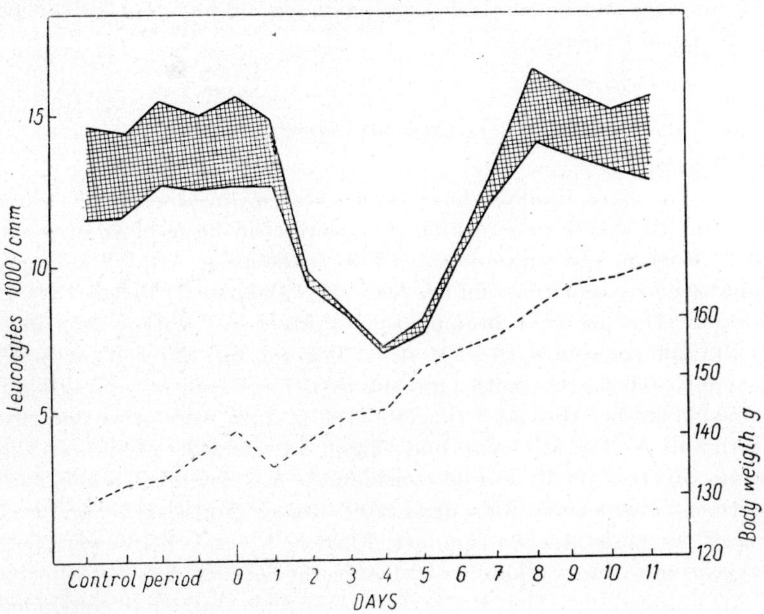

FIG. 2. The changes of white-blood-cells count and body weight (— — — —) after single injection of Vinblastine crystalline sulphate 0·05 mg.

a virtual disappearance of granulocytes accompanies the fall in the W.B.C. it should be noted lymphocytes are also markedly affected. Platelets and R.B.C.'s are well preserved. With smaller doses of VLB a number of animals may fail to show such a response but with larger doses practically all animals show such depression of the blood count. Essentially the same effect is noted if VLB is administered by other routes, even orally, provided adequate dosage is given. The bone marrow shows a parallel loss of leucocyte-forming cells. Comparable VCR injections in rats usually show less consistent and less severe changes. The occasional animals may show a marked alteration, but doses causing this closely approximate lethal levels. Recovery is prompt and is complete within a week, a marked rebound in the W.B.C. is usual, with the differential count showing a predominance of granulocytes. Single injections given repeatedly at 7–10 day intervals elicit a series of typical responses without evidence of exhaustion of the bone marrow or cumulative effects of the alkaloid. The haematological changes in the rat would appear to be similar to those which occur in the human. Since these side effects with VLB are the serious limitation to treatment with increased dosage many attempts have been made to selectively reduce or prevent the blood changes. Treatment with agents such as cortisone, folic acid derivatives and batyl alcohol and chimaera oil have not influenced the haematological response to VLB.

ANTI-TUMOUR ACTION

In vivo. The vinca alkaloids have been tested against numerous mouse and rat tumours by standard screening procedures and have been reviewed recently.[(5)] With mouse tumours the P1534 leukemia in DBA/2 mice proved to be particularly sensitive and has been used extensively by the Lilly group of workers. The per cent prolongation of life over controls was marked. From 100–200 per cent with VCR at 0·2–0·25 mg/kg; 130–150 per cent with VLB at 0·45–0·6 mg/kg; with vinrosidine 127 per cent at 5·0 mg/kg; with vinleurosine 32 per cent at 10 mg/kg. Daily doses were given for 10 days and with both VCR and vinrosidine a significant number of animals showed indefinite survival. VCR showed considerable activity, even when the start of treatment was delayed for 6 days after tumour implantation and resulted in "cured" animals. Ascites tumours, Ehrlich, Freund, S-180 were particularly sensitive to most alkaloids, although the Ehrlich was not affected by VCR or vinrosidine. The Ridgeway osteogenic sarcoma responded to VCR and vinrosidine, but not to VLB or vinleurosine. The L1210 leukemia in DBA/2 mice was resistant to all four alkaloids. Our own more limited experience has been essentially similar. The L1210 leukemia when carried in F1 hybrids, however, has shown a moderate and consistent response to VLB as has an AKR leukemia.[13] The Ehrlich ascites tumour shows little response to VLB when the alkaloid is injected subcutaneously rather than by the intraperitoneal route. Similarly, the tumour if planted subcutaneously is relatively resistant to therapy. This tumour has, however, in our hands yielded interesting information regarding the most effective dosage of VLB. The results have been shown in three separate tables of experiments performed at different times, but with slightly different techniques. In our earliest studies, shown in Table 5, mice were innoculated with undiluted ascitic fluid and because of this large innoculum showed a greater degree of resistance to therapy. All data in each table, however, is comparable. In Table 6 and Table 7 the cells were washed and diluted with saline and counts made of the number of cells innoculated. In general with a smaller number of cells the response to VLB was increased. Table 5 shows that intraperitoneal treatment prior to tumour innoculation had little effect on survival. Previously it was believed that division of the total dose of VLB into aliquots given hourly enhanced activity. With the Ehrlich tumour, however, little evidence of this could be found when doses were given one day or continued into the second day. Division of the dose,

TABLE 5

EFFECTS OF VARIED DOSAGE AND TIMING ON EHRLICH ASCITES TUMOUR—VLB

Dose daily mg/kg	Dose regimen	No. mice	% prolongation of life
1·0	Daily — 2 days prior innoc.	6	0
1·0	Daily — 1 day prior innoc.	5	+21
1·0	Daily — 30 min prior innoc.	8	+ 7
1·0	Simultaneously with innoc.	5	−14
2·0	Day 1 — single dose	7	−16
2·0	Day 1 — hourly × 5	8	+18
1·0	Day 1 — single dose	7	−16
1·0	Day 1 — hourly × 5	7	− 4
1·0	Days 1 & 2 — single doses	4	+40 S
1·0	Days 1 & 2 — hourly × 5	5	+47
0·5	Days 1 & 2 — single doses	6	+54
0·5	Days 1 & 2 — hourly × 5	6	+33
0·25	Day 0 — hourly × 9	6	0
0·25	Day 1 — hourly × 9	6	− 7
0·25	Days 1 & 2 — hourly × 9	5	+35

S = 1 survivor.

however, did not reduce its effectiveness. In Table 6 the increased effect of single doses when extended for 2 or 3 days is shown; little improvement was noted if they were continued for 10 days. The inactivity of subcutaneous injections of VLB is shown. When larger intravenous doses of VLB were

TABLE 6

EFFECT OF VLB AND VCR ON EHRLICH ASCITES TUMOUR

Dose daily mg/kg	Route	Treatment on Day 1	Days 1 & 2	Days 1, 2 & 3	Days 1—10
		% prolongation of life			
VLB					
1·0	i.p.		+200 (5) S	+180 (5) S	
0·5	i.p.	+25 (9)	+160 (6)	+190 (6)	
0·45	i.p.	0 (7)			+170 (6)S
0.35	i.p.			+120 (6)	
0·2	i.p.			+ 48 (6)	
0·1	i.p.			+ 28 (6)	
0·45	s.c.				0 (6)
15·0	i.v.	+ 4 (8)			
10·0	i.v.		— 50 (6)		
10·0 and 5·0	i.v.		+ 30 (6)		
VCR					
1·0	i.p.			— 20 (6)	
0·5	i.p.			+ 61 (5)	
0·2	i.p.			+ 58 (5)	
0·1	i.p.			+ 14 (4)	
VLB 0·1 mg/kg + VCR 0·1 mg/kg	i.p. i.p.			+ 21 (6) S	

Number of mice in brackets. S = 1 survivor.
Tumour innoculum approx. 15 million washed cells.

used some activity was noted, but repeated doses proved toxic. VCR in these tests showed activity when administered for 3 days, but the addition of VLB did not add to effectiveness. Johnson has also commented on the lack of potentiation between the vinca alkaloids, with the possible exception of vinrosidine.[(5)] In Table 7 the tumour innoculum was reduced to 5·8 and 2·5 million cells with a marked increase in the response to treatment. Using varied spacing of three daily doses VLB was highly active when given on days 1, 4 and 7 or when commenced on day 2. It was believed the greater action was related to the opportunity of the mice to recover from toxicity with such dose spacing.

The response of two rat leukemias to vinca alkaloids has been studied extensively in our laboratory. The IRC ascites carried in Fischer rats which is resistant to treatment, and a chloroleukemia carried in Sprague–Dawley rats by intraperitoneal implantation which is extremely sensitive. As

TABLE 7

EHRLICH ASCITES TUMOUR AND DOSE SPACING OF VLB

Dose daily mg/kg	Treatment days	5·8 million cells innoculum			2·5 million cells innoculum		
		No. mice	% prolongation	Survivors	No. mice	% prolongation	Survivors
1·0	1, 2, 3	5	—	5	6	—	5
1·0	1, 3, 5	6	—	6	6	—	6
1·0	1, 4, 7	6	—	3	6	—	5
1·0	2, 3, 4				6	—	4
1·0	4, 5, 6				5	+ 81	0
1·0	6, 7, 8				6	+ 32	0
0·5	1, 2, 3	5	+130	1	6	—	4
0·2	1, 2, 3	5	+ 76	0	5	+106	0
0·1	1, 2, 3				6	+ 47	0

previously noted, rats bearing the IRC tumour when treated with VLB show only a small increased survival. Despite this, such treatment was apparently followed by a dramatic clearing of circulating tumour cells from the blood stream, and histological evidence suggested that large masses of tumour cells in the liver had undergone some form of dissolution. These *in vivo* results are in striking contrast to the apparent extreme sensitivity of IRC cells to VLB in an *in vitro* system to be described later. The chloroleukemia is the most sensitive tumour which we have examined. In Table 8 it may be seen that VLB increases survival and "cured" rats may be readily produced. The tumour also appears sensitive to VCR. The possible use of a rat rhabdomyosarcoma for VLB studies was suggested by an early observation of the dramatic response to VLB of a rhabdomyosarcoma in a child.[14] Publications by Gilman[15] indicated that transplantable rhabdomyosarcomas could readily be produced in rats by the intramuscular injection of nickel sulphide. Miss Capstick[16] has now tested both VLB and VCR in a number of early transplant generations of such induced tumours, but the response has not been dramatic. Whereas the onset of growth after transplantation may be delayed by treatment, little effect of the alkaloids could be demonstrated on growing tumours of this type.

The *in vitro* response to the ascitic forms of the Ehrlich and IRC tumours has been studied by Miss Hawryluk. In these experiments cells have been maintained by usual culture techniques for periods of 48 hr. At intervals cells have been stained to indicate viability, but have also been innoculated into suitable recipients. Although these studies are still in progress, some interesting points have become apparent. The Ehrlich tumour cells when exposed (100,000 per ml) to VLB have proved to be highly resistant to its

TABLE 8

EFFECT OF VLB AND VCR ON CHLOROLEUKEMIA IN SPRAGUE-DAWLEY RATS

	Daily dose mg/kg	Treatment on days	No. rats	Indefinite survivors
VLB	0·07	1 to 10	6	16%
	0·05	1 to 10	6	66%
	0·10	1 to 10	6	100%
	0·15	1 to 10	5	100%
	0·20	1 to 10	4	100%
	0·1	3 to 13	4	100%
	0·2	1	5	0%
	0·2	1 to 5	6	30%
	0·2	5 to 15	4	100%
	0·2	7 to 17	4	100%
	0·2	14 to 24	4	0%
VCR	0·05	1 to 10	6	50%
	0·08	1 to 5	5	80%
	0·25	1 to 5	7	71%

Average survival, 38 controls 22·3 days (17 to 26 days), no indefinite survivors.

action. It has required doses of 0·2 mg/ml to cause cell death with a failure of them to transplant successfully. In contrast, IRC cells under comparable conditions show over 80 per cent cell death with concentrations of 0·00002 mg/ml and a failure of cells to initiate tumour growth at 0·02 mg/ml. These findings are in striking contrast to the *in vivo* studies where the reverse sensitivity was noted. *In vitro* studies have also shown the importance of the time of exposure of cells to the alkaloid. Ehrlich cells, incubated even at maximum dose levels of VLB, are relatively unaffected after 6 hr and readily produce tumours when innoculated into mice. After 48 hr exposure, however, the cells are apparently dead and do not induce tumours. The same time relationship has been shown with IRC cells. The explanation of these *in vitro* findings is not clear at present, but suggests that either cells will recover after some hours exposure to VLB or that the alkaloid may only enter the cell metabolism at a critical stage in cell division, so that it is necessary to have a more prolonged incubation to allow this stage to be reached by the majority of cells.

In this connection, studies by others have indicated that all the active vinca alkaloids are capable to some extent of arresting mitosis of cells when administered both *in vitro* and *in vivo*.[17–21] Arrested metaphase was originally noted in the bone marrow biopsies of a patient treated with VLB. Although metaphase arrest in tumours may be a striking feature, there seems little evidence to indicate a parallelism between this effect and the sensitivity of the tumour *in vivo* to treatment by VLB. Johnson has noted that a number of amino acids may reverse or block such effects of VLB and may actually confer protection of the host against toxic side effects. Aspartic acid allowed host protection, but did not interfere with the antitumour effect of VLB, and has led to some clinical studies with amino acids.[5, 19]

Clinical use of alkaloids

The numerous clinical publications on the response of tumours and leukemias of humans to both VLB and VCR sulphates have been recently reviewed *in extenso*.[5] Only a few points of interest arising from these will be mentioned here. Both the alkaloids have induced a dramatic remission on occasions in a wide variety of tumours, including carcinoma of the lung, breast, ovary, kidney and cervix, chorioepithelioma, rhabdomyosarcoma, reticulum cell sarcoma and neuroblastoma. Carcinomas of the stomach, colon and rectum; melanoma, teratoma, monocytic leukemia, erythroleukemia, astrocytoma and mycosis fungoides have also responded. VLB has been of particular use in Hodgkin's disease, chorioepithelioma, and more recently in breast and bronchogenic carcinoma. VCR, in contrast to VLB, has shown its greatest activity in acute leukemia in children and some lymphosarcomas, but has also been effective in carcinoma of the cervix and possibly has shown a greater consistency of action in a wider spectrum of tumours than VLB. Although the response to therapy may be rapid and dramatic with the virtual disappearance of the lesions, particularly in Hodgkin's disease, lymphosarcoma and acute leukemia, recurrence occurs despite continued maintenance therapy. The duration of remissions varies considerably and is probably longest in Hodgkin's disease and lymphosarcoma and relatively short in leukemias and some solid tumours. When maintenance therapy has not been used, tumours which show regrowth may respond to an additional course of therapy. Many of the tumours in patients treated with vinca alkaloids have previously become refractory to radiation or one or more oncolytic agents. No suggestion of cross resistance has been noted. Curiously enough, despite the close chemical similarity of VLB and VCR, the development of tumour resistance to one alkaloid has not prevented a response to the other.

Although many schedules of dosage have been described, in most cases the optimum tolerated dose is established and then given intravenously as a single dose at weekly or 10-day intervals. Recovery of any depression of the W.B.C. between doses is usually rapid and complete. The tolerance to VCR is considerably less than VLB. In the former case doses of 0·01 to 0·03 mg/kg for a week may be effective. Children with acute leukemia show a better tolerance and may require 0·1 mg/kg for optimal therapeutic effects. The tolerance of VLB is approximately 3 to 10 times that of VCR in adults, but considerable variation exists. VLB is also effective after oral administration, but larger doses are required.

The major toxic manifestations appear to be different and selective for the two chemically similar alkaloids. Neither compound seriously affects haemoglobin or platelet production. VLB, however, may profoundly depress white blood cell forming elements so that intense leukopenia is the limiting factor to increasing dosage in therapy. VCR seldom produces changes in the W.B.C. but neuromuscular toxicity has limited the dosage to be employed. (The apparently increased tolerance of children to the toxic effects of VCR has been noted.) Epilation in 10–20 per cent of cases may follow the use of either drug, with VLB being somewhat less active. Both drugs cause an intense local tissue reaction if not administered intravenously.

MECHANISM OF ACTION

Attempts to demonstrate *in vitro* actions of the vinca alkaloids on tumour cells have been disappointing in the hands of the Lilly group and ourselves.[5, 22] The rapid and profound manner in which tumours may be effected encouraged us to expect readily detectable changes. In general, however, it has been found that VLB and VCR in physiological concentrations (up to 50 μg/ml) showed little effect on respiration and glycolysis of ascites tumour cells. The effects of the alkaloids on cell mitosis suggested studies on nucleic acid synthesis. Again the alterations on *in vitro* tumour systems have not been dramatic, nor have the effects on protein synthesis. *In vivo* studies by Beer, however, originally showed a marked suppression of ^{14}C-formate incorporation in DNA, RNA and DNA bases in bone marrow, when VLB was administered 15 hr previous to the tracer. In contrast, at 2 hr there was little inhibition. Johnson has also stressed the time factor in tissue culture experiments so that *in vitro* systems employing short incubation times might not be responsive. The specific nature of the *in vivo* action of VLB was emphasized by Beer, since rapidly dividing cells such as those of regenerating liver were not affected, nor was their DNA and RNA synthesis. In similar experiments VLB did not affect nucleic acid formation in Ehrlich ascites cells. This tumour, however, has since proven to be very unresponsive to VLB when the alkaloid is given subcutaneously, as occurred in this experiment.

Very recently Drs. Richards, Jones and Beer have extended the *in vitro* studies. VLB inhibited the incorporation of isotope into the nucleic acids of washed cell suspensions of rat bone marrow and thymus when they were incubated with ^{14}C-formate. The effect was more marked on DNA than RNA and the specific activities of the adenine and guanine of the DNA isolated from the thymus cells were equally depressed. The thymine labelling was also decreased but to a somewhat lesser extent than the purines. The alkaloid also affected these bases in the "acid soluble" portion of the incubated thymus cell suspension. Whereas the labelling of adenine and guanine was decreased, that of hypoxanthine and thymine was increased. In addition there was an accumulation of radioactive serine in the presence of VLB.[(23)] This suggested that the alkaloid might be blocking some step in purine synthesis or incorporation into the nucleic acids at a point subsequent to the formation of inosinic acid.

METABOLISM OF THE VINCA ALKALOIDS

Information on the rate and route of excretion of the alkaloids and on their metabolism has been sought to aid in the study of their mode of action and to possibly suggest a means of enhancing their activity without increasing toxicity. Beer originally used a semi-quantitative colorimetric method and found that VLB given intravenously rapidly disappeared from the blood and only small amounts of the unchanged alkaloid appeared in the urine.[(22)] More recent studies have attempted to prepare ^{14}C-labelled alkaloids by extracting plants grown in an atmosphere of radioactive CO_2. At present sufficient ^{14}C-labelled VLB has not been isolated for biological experiments. VLB tritiated by the Wilzbach method has more recently been studied extensively in this laboratory. Certain difficulties of a chemical nature using this method have now been resolved by Dr Beer. The formation of appreciable (up to 50 per cent) amounts of dihydro-VLB by hydrogenation in the tritiation exposure process has led to contamination of the end product which was not recognizible by chemical methods and was not separable from VLB by any of the recrystallization and chromatographic methods previously found satisfactory for the preparation of pure VLB. An effective separation of the tritiated dihydro compound and VLB has now been developed, however, using chromatography on cellulose phosphate.[(24)] Samples of the two pure tritiated alkaloids have therefore become available for biological studies.

The intravenous injection in the rat of pure tritiated VLB was followed in 26 hr by the excretion in the urine of only approximately 5 per cent of the dose, measured as non-volatile radioactivity. Of this, 2 per cent appeared in the first 3 hr. Heart blood at the end of 26 hr contained only 0·08 per cent of the injected radioactivity in 1 ml. The total radioactivity of the blood at this time was approximately 1·2 per cent of the injected dose. In view of the absence of other than small amounts of the alkaloid or metabolites in the urine or blood, the fate of the compound in the animal was investigated. In further studies VLB was again found to be removed rapidly from the blood stream — 1 to 1·5 per cent of the dose radioactivity was circulating 30 min after intravenous injection. No tissue apart from the liver showed a concentration of radioactivity, including bone marrow and brain, when measured at 24 hr. At this time 20–30 per cent of the dose radioactivity was found in the gut contents. Further studies using rats with a bile canula indicated that 20 per cent of activity was eliminated in the bile in 24 hr and about 12 per cent in 4 hr. Not all radioactivity reached the gut from bile. Unchanged VLB accounted for less than 1/10 of the bile radioactivity so that less than 2 per cent of the alkaloid is excreted in 24 hr. These experiments suggest that the injected alkaloid may be rapidly trapped in the liver and there be converted to metabolites, some of which pass into the bile.[24]

COMPOUNDS RELATED TO THE VINCA ALKALOIDS

In their chemical studies on *Vinca rosea* the Lilly Group have isolated a remarkable number of compounds from the plant. Besides the four active compounds previously discussed, at least twelve related compounds and ten monomeric alkaloids have also been isolated and identified. In addition, various derivatives of vinleurosine and VLB have been prepared.[5] These compounds had been originally screened for biological activity at the time of isolation. With the increasing data indicating that highly specific types of activity and toxicity may be associated with compounds with only minor chemical differences, as has been emphasized in this paper, attention has recently been focused on the various related compounds. At present Mr. Galliford in our laboratory is testing these compounds to see if they possess any type of biological action or if they possibly can compete or block certain activities of the active alkaloids. These include VLB derivatives obtained by Dr Beer from modified alumina columns or biological metabolites, and monomeric and other related alkaloids supplied from the Lilly Laboratories. Particular interest at present has centred on the dihydro VLB which, as mentioned previously, is separable from VLB mixtures only with difficulty. This compound has proven to have a relatively low toxicity when administered by a single intraperitoneal injection to mice — groups of three animals each tolerating doses of 30, 60 and 90 mg/kg without mortality. A dose of 5 mg/kg of VLB similarly tested yields a mortality of approximately 50 per cent. In the rat intravenous doses of 20 mg/kg have been tolerated, although the animals showed a fall in W.B.C. similar to

VLB. In view of this unexpected relative absence of toxicity a number of mice bearing Ehrlich ascites tumours were treated for 3 days with 5, 10 and 20 mg/kg daily. Whereas the lowest doses showed an increased survival time of 12 per cent, with the highest dose three of six mice are still surviving some three times the length of controls. Although these experiments are limited because of the supply of material, they suggest that this compound possesses a relatively greater anti-tumour effect than toxic action, and if this is so it may be a preferable therapeutic agent to VLB.

ACKNOWLEDGEMENTS

This paper reviews contributions made in the laboratory by Drs. C. T. Beer, J. F. Richards, F. Lotz and Misses A. Hawryluk and V. Capstick, with the technical assistance of Mrs. D. King, J. Bell, M. Wilson, D. Gaffney, Mr. A. Renner and Mr. B. Galliford. Financial support for this work has been gratefully received from the National Cancer Institute of Canada, Medical Research Council of Canada and the Eli Lilly Company of Indianapolis.

REFERENCES

1. SVOBODA, G. H., JOHNSON, I. S., GORMAN, M. and NEUSS, N. *J. Pharm. Sci.* **51** 707 (1962).
2. PECKHOLT, T. *Ber. Deutsch. Pharm. Ges.* **20**, 36 (1910).
3. HUGH-JONES, P. *Lancet,* **2**, 891 (1955).
4. WATT, J. M. and BREYER-BRANDWIJK, M. L. *The Medicinal and Poisonous Plants of Southern and Eastern Africa,* E. & S. Livingston Ltd., London (1962).
5. JOHNSON, I. S., ARMSTRONG, J. L., GORMAN, M. and BURNETT, J. P. JR. *Cancer Research,* **23**, 1390 (1963).
6. NYE, L. J. J., and FITZGERALD, M. E. *Med. J. Australia,* **2**, 626 (1930).
7. CORKHILL, A. B. *Med. J. Australia,* **1**, 86 (1929).
8. LEE, D. K. H. and DREW, W. R. M. *Med. J. Australia,* **1**, 699 (1929).
9. NOBLE, R. L., BEER, C. T., and CUTTS, J. H. *Assoc. New York Acad. Sci.* **76,** 882 (1958).
10. JOHNSON, I. S., WRIGHT, H. F. and SVOBODA, G. H. *J. Lab. Clin. Med.* **54,** 830 (1959).
11. NOBLE, R. L. *Proc. Canad. Cancer Res. Conf.* **4,** 333 (1961).
12. ARMSTRONG, J. G., personal communication.
13. CUTTS, J. H., BEER, C. T. and NOBLE, R. L. *Cancer Res.* **20,** 1023 (1960).
14. WHITELAW, D. M. and TEASDALE, J. M. *Canad. M.A.J.* **85,** 584 (1961).
15. GILMAN, J. P. W. *Cancer Res.* **22,** 158 (1962).
16. NOBLE, R. L. and CAPSTICK, V. *Proc. Amer. Assoc. Cancer Res.* **4,** 48 (1963).
17. CUTTS, J. H. *Cancer Res.* **21,** 168 (1961).
18. CARDINALI, G. and ENSIN, M. A. *Blood,* **21,** 102 (1963).
19. JOHNSON, I. S., VLANTIS, J., MATTAS, B. and WRIGHT, H. F. *Proc. Canad. Cancer Res. Conf.* **4,** 339 (1961).
20. VAITKEVICIUS, V. K., TALLEY, R. W., TUCKER, J. L. and BRENNAN, M. J. *Cancer,* **15,** 294 (1962).

21. Walker, D. G. and Wright, J. C. *Cancer Res.* **22**, 1267 (1962).
22. Beer, C. T. *Proc. Canad. Cancer Res. Conf.* **4**, 355 (1961).
23. Richards, J. F., Jones, R. G. W. and Beer, C. T. *Proc. Amer. Assoc. Cancer Res.* **4**, 57 (1963).
24. Beer, C. T., Wilson, M. L. and Bell, J. *Canad. J. Physiol. Pharmacol.* **42**, 1 (1964).

HISTORY OF *RAUWOLFIA SERPENTINA* AND OUR EARLY INVESTIGATIONS

B. MUKERJI

Chittaranjan National Cancer Research Centre, Calcutta, India

THOUGH many important drugs are described in two ancient Indian texts of *Charaka* and *Susruta* (*ca.* 100–400 A.C.) which list nearly 2000 items with diverse therapeutic claims, *Rauwolfia serpentina* did not find any mention there. The earliest authentic record appears to be in 1563 by Gracia al Horto.

In spite of the fact that he described *Rauwolfia serpentina* Benth. as the "foremost and most praiseworthy Indian Medicine", no substantial scientific investigation appears to have been carried out till Bose and Sen[(1)] reported their clinical findings with crude preparations of this drug in 1931. Closely following this, Siddiqui and Siddiqui[(2)] in the same year isolated for the first time five crystalline alkaloids, e.g. ajmaline, ajmalinine, ajmalicine, serpentine and serpentinine.

The first pharmacological study on modern lines was undertaken at the Department of Pharmacology of the School of Tropical Medicine early in 1932 and our joint findings were reported by Chopra, Gupta and Mukerji[(3)] in 1933 in the *Indian Journal of Medical Research*. In this paper the action of an alkaloid with a m. p. 135°, isolated in the Department of Chemistry of the same Institution, was reported. The workers presumed that the substance used in their experiments corresponded to "ajmaline" of Siddiqui. No direct comparison of the two substances was possible at the Chemistry Laboratories of the School of Tropical Medicine, because of the non-availability in sufficient quantities of the alkaloids isolated by Siddiqui. The conclusions made by the workers on the basis of the pharmacologic evaluation methods then available are recorded below:

> The alkaloid from *Rauwolfia serpentina* has a stimulant effect on the plain muscle of the intestine and uterus.
>
> The systemic blood pressure falls due to dilatation of the blood vessels of the splanchnic area.
>
> The alkaloid has a pronounced effect on the central nervous system. In sublethal doses injected into the lymph sac of frogs, narcosis quickly ensues. In mammals the alkaloid produces symptoms which are attributable to its depressant effect on the various cerebral centres in the reverse order of their development. There is also some evidence to indicate that there is some depression of peripheral nerve cells of the body.

The alkaloid on account of its pronounced cerebral depression properties should prove to be a valuable sedative drug. It lowers the blood pressure and if administered in proper dosage should be of value as a remedy against hypertensis. The drug is likely to be a valuable addition to the armamentarium of physicians and further work to place it on a definite therapeutic basis is in progress.

Professor Chopra was a pharmacologist with a strong clinical bent of mind and as soon as the acute toxicity studies with the *Rauwolfia* alkaloid were completed in the laboratory and a dosage level safe enough for human administration could be worked out, he started out with human trials with both the standardized crude extract and the crude powdered materials from *Rauwolfia*. On the patients under his care at the Carmichael Hospital for Tropical Diseases, Professor Chopra and Dr Gupta obtained satisfactory correlation of the data obtained in the Pharmacology Laboratory and he was so impressed with particularly the prolonged sedation effect of the drug, that he undertook, in collaboration with the Director of the Ranchi Mental Hospital in Bihar, a controlled clinical trial of the drug in several types of mental disorders showing violent maniacal symptoms. Results obtained in some cases of schizophrenia were encouraging but uniform recordable data which could stand the test of critical analysis were not available. In one set of trials, actually a recrudescence of violent symptoms with increased agitation and excitement was recorded. In view of the inconclusive nature of the data, Professor Chopra discouraged the publication of the data obtained from these human trials at that time.

Meanwhile, clinical trials with the *Rauwolfia* drugs were confirmed outside the Hospital and through his influence Professor Chopra was able to get reports of these cases treated by private physicians outside the regular hospital clinics. An overall analysis of the data obtained from these sources left little room for doubt that *Rauwolfia* drug was a definite contribution to therapy, particularly in cases with high blood pressure and mental agitation symptoms. Guided by these observations, he advised me to concentrate all my energies on the isolation of the active principle of the drug in collaboration with the chemistry group at the School of Tropical Medicine and undertake repetitive pharmacological experiments to clarify further the nature and seat of action of the drug. In late 1933, through the help of the Rockefeller Foundation, Professor Chopra arranged to depute me to the Peiping Union Medical College in China where successful work on the isolation of ephedrine from the Chinese drug *Ma Huang* was already accomplished. It was felt that my acquaintance with the methodology employed in such investigations on plant products would be helpful in advancing further our studies on the Indian drug *Rauwolfia*. When I reached Peiping, however, the group of pharmacologists headed by Read, Chen and Schmidt had moved out from there to other parts of the world and interest in plant products and their pharmacological evaluation remained in abeyance for the time being.

Returning to India in 1937 after completion of my pharmacology studies in America, I rejoined my duties and Professor Chopra's first advice was to return to the *Rauwolfia* studies as soon as possible, and he further arranged to get several collaborators in his laboratory to attack the problem from different angles. On the impression that sedation produced by the *Rauwolfia* drug might have a peripheral component also, we studied the action of ajmaline, then obtainable in a pure form from Dr Siddiqui, on nerve impulses employing isolated nerve muscle preparation of frogs. The alkaloid was found to diminish the discharge frequency of nerve impulses. In 1941, Chopra and Chakravarti[5] made a comparative study of the three isolated alkaloids of the plant, namely, ajmaline, serpentine and serpentinine. In decerebrate animals ajmaline raised the blood pressure while in spinal animals there was fall, which was more marked if the blood pressure was previously raised through adrenaline infusion. They suggested that ajmaline was a central vasomotor stimulant but had depressant effect on the heart and peripheral vessels. The sum total of this effect was a rise of blood pressure in decerebrate animals because the vasomotor stimulation masked the peripheral depressant effect. When, however, the vasomotor centre was excluded, as in spinal preparation, the peripheral effect alone prevailed. The effect of this product on intact animals was not reported.

In 1942 Chopra, Bose, Gupta and Chopra[6] obtained the three alkaloids from Siddiqui and investigated the comparative action of these alkaloids. They confined themselves to a study of the effect on blood pressure. They concluded that the chief pressure-reducing factor in the plant was serpentine. Although all the three alkaloids produced a fall in blood pressure after experimental hypertension, ajmaline and serpentine raised the blood pressure in intact normal animals. The crude extract also produced a fall in blood pressure. The hydrochloride of the total alkaloids from which ajmaline, serpentine and serpentinine had been removed had no effect on blood pressure. The alkaloids had both direct effect on the plain muscle of blood vessels and through the vasomotor control of the blood vessels. These results were more or less a confirmation of their earlier observations in 1941. At this stage it appeared that ajmaline and serpentinine belonged to one group in respect of their effect on blood pressure and intestine muscle while serpentine to the other. On the other hand, in respect of effect on the C.N.S., ajmaline and serpentine were stimulants and serpentinine was the sedative alkaloid.

In 1943 Chopra, Gupta, Bose and Chopra[7] focused their attention on the hypnotic principle of the plant. As a result of this study they revised the earlier conclusion arrived at by Chopra and Chakravarti regarding the sedative properties of serpentinine. This alkaloid was found to be a C.N.S. stimulant similar to ajmaline and serpentine. Regarding the nature of the hypnotic and sedative principle of the drug, it was suggested that the alcohol-soluble fraction or fractions other than the three isolated alkaloids was responsible for this property of the plant.

In 1944 Gupta, Kahali and Dutta[8] reviewed the whole position relating to the hypnotic principle of *Rauwolfia serpentina*. These workers recorded that the resin fraction, free from all traces of alkaloids, produced a sedative and hypnotic activity in cats, rabbits, frogs and guinea-pigs. The effect of this fraction was similar to that produced by the standard extract, commencing 3–4 hr after administration and lasting for more than 24 hr.

From 1944 onwards, as a result of the Second World War in which India was involved and also because of our failure to come to any conclusion regarding the active component or components of the *Rauwolfia* drug, the problem could not be pursued further. Meanwhile, Professor Chopra retired and both Dr. Gupta and myself, the two other collaborators, had to engage ourselves in other problems pertaining to standardization of Indian crude drugs and finding suitable substitutes for imported drugs from Indian materia medica to help the acute shortage in drug supply to the army, navy and air force employed in the South East Asian theatres of war.

With the termination of the Second World War, around 1946—1947, it was possible again to turn my attention to the *Rauwolfia* problem. Professor Chopra's contacts with Professor Arthur Stoll of Sandoz, Switzerland, made it possible to arouse his interest in the investigation of this Indian plant product and a chemicopharmacological investigation was started in his laboratories around 1948. About this time, some Indian chemists working in the laboratories of Professor Robinson in England also took up independently the chemical isolation and characterisation of the active principles of this drug, which had already earned a reputation in India as a really effective remedy, even though the pharmacological evaluations carried out by us earlier were inconclusive.

In 1950 I had an opportunity of coming out to America and I brought with me a supply of crude *Rauwolfia* powder with a view to get the American pharmacologists and drug research workers interested in finding out the active principle of this Indian drug, which to my mind, and which as a result of my own experience both in the laboratory and in the human patients in the clinics and hospitals, was a sovereign remedy both for blood pressure and for its sedative action on the central nervous system. At this time the word "tranquillizer" was not known. I submitted the material for chemical, pharmacological and clinical evaluation to the well-known drug house Smith Kline and French at Philadelphia and worked with their research staff for some time in an effort to prove the efficacy of the material. However, success eluded my efforts, particularly because we were then concentrating on the isolation of the *strong* bases from the alkaloid fractions and were not mindful of the *weak* bases present in the water-soluble resinous fractions of the plant material.

Chemical methods of isolation of very similar active principles through column and paper chromatography and counter-current distribution techniques were not available to Indian workers and we were unable to separate the various fractions in sufficient purity and in sufficient quantities to permit more accurate studies of their pharmacology. Apparently inconsistent data were obtained during the 18 years of our approach to the problem from time to time. This was disappointing but nevertheless did not shake our faith in the efficacy and potentiality of the *Rauwolfia* drug. Professor Chopra's strong clinical bias and his observations on human patients were supported by all his colleagues and co-workers, many of whom had also used the crude drug powder and its alcoholic extract in their own clinical practice.

In 1952, Müller, Schlittler and Bein first reported the isolation of reserpine and thereafter reserpine became the focus of attention of numerous investigations for use in hypertension and mental disorders. Numerous papers have since been published on the pharmacological action and therapeutic uses of this drug and excellent reviews have also been published. It is not necessary, therefore, in this paper to restate all the findings. An interesting development of the discovery of reserpine from the Indian *Rauwolfia* drug which needs emphasis here is that reserpine has opened up a new area of interest in psychopharmacology and neuropharmacology and is being now increasingly used as a tool in diagnostic and psychiatric therapy.

While the discovery of reserpine has undoubtedly ushered in a new field for research in plant products, the problem of the *Rauwolfia* drug does not appear to have been solved yet. We know now the nature of many of the *Rauwolfia* alkaloids and several alkaloids have been well characterized chemically and some have also been synthesized. Numerous semi-synthetic and synthetic derivatives have also been prepared. However, it is not yet clear as to whether reserpine can account for all the different pharmacological activities seen with the crude whole drug powder or with some of its alkaloidal fractions. In 1955 I carried out, with my colleagues at the Central Drug Research Institute at Lucknow, a comparative study of the active principles of the *Rauwolfia* drug with pure reserpine and found evidence that the total alkaloidal fractions of the crude drug were more potent in their tranquillizing action than reserpine alone. Alkaloids other than reserpine (e.g. raunescine, etc.) have been reported possessing a potent biological activity on the central nervous system. Possibly some additive or potentiation effect of the individual alkaloids is responsible for the action of the whole drug. The whole drug also contains some principles which bring about undesirable side actions such as purgation and sexual debility. It is possible, however, to remove some of these components without diminishing the sedative and hypotensive properties inherent in the drug. Recent experiments have also helped in elucidating the fact that the two most pronounced actions of the *Rauwolfia* drug may perhaps be due to two *different active principles* and not due to reserpine alone. Future work would undoubtedly help to clarify some of these complexities which remain unsolved even today. My own conviction is that most plant products available from natural sources contain more than one active ingredient and that the sum-total of action is due to an alkaloidal complex of a macro-molecular nature. This view is also supported by the recent work of Siddiqui from Pakistan regarding "serp-ajmaline".

CONCLUSION

Both the therapeutic virtues of *Rauwolfia serpentina* — its sedative action on the C.N.S. and its blood pressure lowering property — were recorded by early Indian physicians and it is interesting to note that these clinical findings could be supported by laboratory pharmacological investigations nearly 32 years ago by the Indian pharmacologists. These reports, however, did not create sufficient interest in the Western world to enable a better and more critical study of the drug. Because *Rauwolfia* contained a number of similar alkaloids, it presented difficulties in isolation of the most potent active ingredient, reserpine, with methods then available. The Indian pharmacologists and chemists were probably too early on the scene and with inadequate paraphernalia then available in India, more positive results could not be reported. It is a matter of satisfaction to all pharmacologists and clinicians that through the help of modern methodology, the Swiss workers at the CIBA Laboratories have ultimately succeeded in clarifying the problem. Indian *Rauwolfias* have now created a place for themselves in modern medicine, comparable in every way to the Chinese drug *Ma Huang* (*Ephedra vulgaris*) from which ephedrine was first isolated by Chen and Schmidt and which is now incorporated in every pharmacopoeia of the world.

All ancient civilizations of China, India, Egypt, Peru, etc., contain records of a rich materia medica from which diligent search, using modern pharmacological laboratory and clinical studies, is likely to yield new worth-while remedies for the alleviation and treatment of many human maladies for which no useful cures are yet known, e.g. rheumatic and other metabolic diseases, cancer and leukemia, pediatric and geriatric conditions not amenable to treatment, etc., etc. The attention of modern pharmacologists should be drawn to this rich storehouse of knowledge so that more investigations are concentrated in this field.

REFERENCES

1. Sen, G. and Bose, K. C. *Indian Med. World,* **2**, 194 (1931).
2. Siddiqui, S. and Siddiqui, R. H. *J. Indian Chem. Soc.* **8**, 667 (1931); **9**, 539 (1932)

3. Chopra, R. N., Gupta, J. C. and Mukerji, B. *Indian J. Med. Res.* **21**, 261 (1933).
4. Chopra, R. N., Das, N., and Mukerji, B. *Indian J. Med. Res.* **24**, 1125 (1937).
5. Chopra, R. N. and Chakravarti, M. D. *Indian J. Med. Res.* **29**, 763 (1941).
6. Chopra, R. N., Bose, B. C., Gupta, J. C. and Chopra, I. C. *Indian J. Med. Res.* **30**, 319 (1942).
7. Chopra, R. N., Bose, B. C. and Gupta, J. C. *Indian J. Med. Res.* **31**, 41 (1943).
8. Gupta, J. C., Kahali, B. S. and Datta, A. T. *Indian J. Med. Res.* **32**, 183 (1944).
9. Kohli, J. D. and Mukerji, B. *Proc. Symposium on Indigenous Drugs and Insecticides,* Nat. Inst. Sci. India Bull. No. 4, 1955.
10. Müller, J. M., Schlittler, E. and Bein, H. J. *Experientia,* **8,** 338 (1952).

THE PHARMACOLOGY OF *RAUWOLFIA* ALKALOIDS

H. J. Bein

Research Laboratories of the Pharmaceutical Department of CIBA Limited Basle, Switzerland

In presenting a short review of the pharmacology of *Rauwolfia* it is impossible to deal with every aspect of the pharmacology of this plant family and the pharmacological actions of the dozens of alkaloids which have so far been isolated. I therefore propose to confine my remarks to the pharmacology of the principal alkaloid, reserpine, and to deal mainly with its actions on the sympathetic nervous system. Although a wealth of pharmacological information on reserpine has been accumulated since its isolation 11 years ago (Müller, Schlittler and Bein [36a]) the mechanism of its antihypertensive and its sedative effects have still not been fully elucidated.

Influenced by the experiments and physiological concepts of Hess, we suggested that reserpine acts on central integrating mechanisms; a reduction in the function of central sympathetic substrates appeared to us to be one of the predominant features presented by the complex pharmacological picture of reserpine.[6, 11] However, it was not possible to attribute every component of the action of reserpine to a known functional entity within the autonomic nervous system. This was particularly true of its action on the intestinal tract which we later explained by reference to the liberation of 5-HT.[8] Although the sites of attack of reserpine appeared to be predominantly centrally located, peripheral sites had to be taken into consideration as well,[11] especially in the light of results yielded by experiments on isolated vessels (Tripod and Meier).[56] The very prolonged pharmacological action of reserpine and the pronounced latency period preceding its onset of effect led to the assumption that reserpine might possibly not act *per se* (Bein).[6] This idea has received additional support from the finding that, at the height of its gradually unfolding pharmacological effect, reserpine could no longer be demonstrated in the tissues by fluorometric methods of analysis. Using labelled reserpine, Sheppard *et al.*[51] and Plummer *et al.*[43], however, have convincingly shown that, even at low pharmacologically active doses, reserpine is detectable in the tissues throughout the period of demonstrable response to the drug. The available data on the metabolism, fate, and excretion of reserpine no longer support the view that reserpine must necessarily act by secondary mechanisms.

The observation of Pletscher, Shore and Brodie[42] in 1955 that reserpine causes a loss of 5-HT from brain tissue served not only to spark off fresh investigations on the action of reserpine and on the function of amines in the central nervous system, but also made it appear likely that secondary mechanisms were in fact involved. It was Vogt[57] who first found that the concentration of an endogenous amine, namely noradrenaline, in the brain could be altered by drugs; these drugs were compounds which stimulate the centres controlling the sympathetic system. In 1956 Holzbauer and Vogt[28] reported that reserpine reduced the noradrenaline concentration in the cat's brain and added that "other tissues containing noradrenaline might be equally affected". It was these authors who thus pointed the way to what has since probably proved the most promising approach to an understanding of certain effects produced by reserpine.

Administration of reserpine causes a lowering of 5-HT and noradrenaline levels in various tissues, although marked differences exist in the response of these tissues: for example, the 5-HT levels in the brain are lowered by single doses which have little effect on platelet or intestinal 5-HT. With regard to noradrenaline, Carlsson and Hillarp,[19] who were the first to note a marked fall in the catecholamine content of the heart, found that the heart was extremely sensitive to reserpine; the brain was next in order of sensitivity, whereas the adrenals proved most resistant to the action of reserpine.

Although there are differences in the response displayed by various organs to the amine-lowering action of reserpine, Brodie *et al.*[15] showed that the effects of reserpine on the brain content of serotonin and catecholamines are similar and that depletion and subsequent repletion of noradrenaline and 5-HT stores occur concomitantly. It therefore seems reasonable to assume the mode of action of reserpine on the stores of the two types of amines is similar. Since the appearance of the original papers, the number of publications dealing with the effect of reserpine on tissue amines has greatly multiplied (see the recent reviews of Shore,[52] and Brodie and Beaven[13]). Stored substrates are not inert materials but are constantly being synthetized and degraded. The amine content of stores is the product of a dynamic balance between the rates of synthesis and degradation. Generally speaking storage must either be attributed to local synthesis or to the operation of a concentrating mechanism. According to Brodie and Beaven,[13] reserpine lowers the content of the stores of both 5-HT and catecholamines by blocking this concentrating or transporting mechanism. Reserpine also interferes with the uptake of catecholamines by other tissues.

It shares this latter property with a variety of other substances (see Ref. 52) which, however, have no analogous actions on the central nervous system. The question as to the importance of the depletion of amines with regard to the mode of action of reserpine and of the relative importance of one amine to another has given rise to much discussion:

When it was first shown that reserpine releases stored serotonin, the hypothesis was advanced that the central actions of the alkaloid are mediated through the free indole derivative (Brodie and Shore[14]). The discovery that reserpine also releases noradrenaline has led to the assumption that it is the depletion of noradrenaline in the brain which determines the central actions of reserpine, or that it is a lack rather than an excess of endogenous amines in the brain in active form which should be regarded as being responsible for the effects of reserpine (Carlsson *et al.*[18]). In 1960 Brodie and co-workers[16] reported that in "cold-stressed animals" reserpine depletes most of the brain noradrenaline but only negligible amounts of the brain 5-HT, although it does not elicit sedation. They therefore assume that the sedative action of reserpine is not related to loss of brain noradrenaline but rather to a change in brain 5-HT. The original observation made by Garattini and Valzelli[24] that rats subjected to cold as well as heat were not sedated by reserpine and their brain 5-HT not lowered could, however, not be confirmed in more recent experiments undertaken by a number of investigators.[1, 30, 33, 59] Nor have we ourselves seen any prevention of the sedative action of reserpine in mice kept at 4°C.

These experiments raise the question of attempting to correlate behavioural effects with changes in the amine concentration in the brain. When I tried to do this on the basis of results reported in the literature, I found the task extremely difficult, at least as regards a very considerable number of papers. This was no doubt partly because of the difficulties encountered in describing and quantitating central effects of a drug, partly because often not enough attention is paid to the importance of the dose of reserpine and the time course of its effect. Sometimes biochemical results obtained in one species and reported in the literature are transferred to another species without due consideration being given to species differences and dosages.

With regard to the temporal course, Sheppard and Zimmerman[49] were the first to state that they could not convince themselves that in the guinea-pigs the gross behavioural effects of medium doses of reserpine (0·1 mg/kg) were correlated with the changes in the brain concentrations of 5-HT and/or noradrenaline[49, 50] (see also Ref. 39). Nor did Walaszek *et al.*[58] find in chicken any correlation between the sedative effect of reserpine and the concentration of 5-HT and noradrenaline in the brain.

The importance of the dosage of reserpine is borne out by von Euler's and Lishajko's[20] observations. According to their recent findings, the action of reserpine on the release of amines, which has been extensively studied *in vitro* and has also been taken as a basis on which to explain phenomena seen *in vivo,* depends very much on the concentration of reserpine. Only very high concentrations of reserpine increase the release of noradrenaline from storage granules prepared from bovine splenic nerves, a phenomenon which these authors therefore regard as unspecific; lower concentrations down to 10^{-8} M of reserpine, however, *inhibit the release of noradrenaline* but do not prevent the uptake of amines.

The situation is further complicated by the fact that the catecholamine stores are probably composed of various compartments having different functions (see Trendelenburg[55]). Experiments with labelled noradrenaline have shown that a form of noradrenaline with a high turnover rate (so-called "available noradrenaline"), which is probably metabolized mainly by catechol-O-methyl-transferase, can be distinguished from a form with a low turnover rate (so-called "bound noradrenaline") which is metabolized mainly by monoamine oxidase. The noradrenaline liberated by reserpine seems to be metabolized chiefly by monoamine oxidase. This suggests that reserpine causes depletion of the stores by acting upon one particular compartment, i.e. on the "bound noradrenaline" (Kopin, Hertting and Gordon,[32] Kopin and Gordon,[31] Potter and Axelrod[44]).

Moreover, there are also conflicting reports as to whether a causal relationship exists between changes in the subcellular distribution of amine concentration and changes in behaviour. Thus, for example, various central nervous depressants (besides reserpine, phenobarbitone, α-methyl-DOPA, chlorpromazine) significantly increased the proportion of free 5-HT regardless of whether they increased, decreased, or had no effect at all on total 5-HT (Schanberg and Giarman[47]). Green and Sawyer[26] could not correlate behavioural changes with any specific distribution pattern of free and stored 5-HT.

Finally, a change in the total amine content of the brain can also be achieved by a variety of agents and procedures without causing sedation: for example, central stimulants,[57] nicotine,[48] Shigella shigae toxin,[35] vincamin,[25] "stress",[21] LSD_{25},[21] and brain lesions[27] — to mention only a few examples.

The use of amino-oxidase inhibitors and of experimental set-ups which exploit interactions between reserpine and other drugs have also failed so far to provide any decisive clues as to the role played by amines in the brain and the mode of action of reserpine, because the actions reported are definitely dependent on the species or on the compound chosen as enzyme inhibitor.[22, 41, 45, 46, 53] There was also no correlation between changes in amine concentration and in the behaviour of the individual animals (Sanan and Vogt).[46]

Further evidence in this field has been provided by the experiments of

Gal, Drewes and Barraclough,[23] who maintained rats on a tryptophan-deficient diet. These animals had a low brain concentration of 5-HT, whereas the concentration of noradrenaline was normal. Reserpine caused a prolonged depletion of 5-HT, whereas the noradrenaline concentration returned to normal much sooner. When a second injection of reserpine was given at this stage, it produced sedation though there was little 5-HT in the brain. These investigators concluded that in the rat the sedative effect of reserpine does not depend on the presence of 5-HT in the brain.

The effect of reserpine on the brain amines has focused so much attention on this peculiar aspect of its mode of action that its effects on other tissue constituents are likely to be neglected. Besides the amine concentration, reserpine is also reported to influence the concentration of glycogen,[2, 3] acetylcholine,[24,40] γ-aminobutyric acid,[4] nucleic acids,[5, 29] and substance P.[54]

Be that as it may, there is ample evidence that reserpine acts selectively on the central nervous mechanisms and that it differs decisively from general central nervous depressants.[7–9] In self-stimulating rats, Olds, Killam and Bach-y- Rita[38] found that reserpine markedly depressed the response rate in animals stimulated in the hypothalamus or in the amygdala, but that it produced only minor depression when the electrodes were located in the septal region. Phenobarbitone, on the other hand, did not show such selective effects in the areas stimulated.

In view of the selectivity of action of reserpine, one is bound to conclude that the different central substrates must vary in sensitivity if the various central components of the action of reserpine are all due to *one* intimate mechanism. The possibility should, however, be considered that the complex pharmacological picture of reserpine is due to more than one basic effect.

Thus, although no causal relationship between depletion of brain amines and sedation in response to reserpine has been firmly established, there seems to be no doubt that reserpine does exert an action on the autonomous nervous system by inducing a loss of neurohumoral transmitter in the sympathetic portion (Carlsson *et al.*;[18] Muscholl and Vogt[37]) and that one of its predominant effects is a diminution in sympathetic tone. In this connection one is tempted to speculate that loss of the neuro-transmitter might not only be the reason for its blocking action on peripheral pathways of the sympathetic system but might also be an indication that similar events may be taking place in the central portions of the sympathetic nervous system — assuming, of course, that it would be possible at all to demonstrate an effect on such central substrates.

Since drugs which stimulate central sympathetic activity cause a fall in the catecholamine content of the brain and of the adrenals, and since the adrenal medulla in the cat loses its amines in response to reserpine only on the innervated side, Vogt concludes — by analogy — that reserpine activates the sympathetic centres. This sympathetic stimulation would be masked by a diminution in function of the peripheral pathways. More recent investigations, however, have shown that reserpine exerts a direct effect on the denervated medulla as well (e.g. Mirkin,[36] Callingham and Mann[17]). As an alternative to the idea that reserpine produces central sympathetic stimulation, one might also consider the possibility that, in response to reserpine, a functioning sympathetic nerve plays a permissive role as regards the liberation of amines in the adrenal.

During the first two to three hours after intravenous injection of high doses of reserpine, we found no impairment of function in peripheral sympathetic pathways, even though the drug's well-known pharmacological effects — such as relaxation of the nictitating membrane, fall of arterial pressure, bradycardia, inhibition of centrally mediated pressor reflexes — were fully apparent. The centrally mediated pressor reflexes in the cat evoked by electrical stimulation of the centripetal vagus or sciatic nerve are transmitted by the sympathetic nervous system. They are blocked by reserpine at a time when stimulation of the efferent splanchnic nerve is fully effective and when no ganglionic blocking activity and no inhibition of postganglionic transmission to the nictitating membrane can be detected.[6, 7, 10] Only after a much longer latency period, usually only after more than 8 hr, is impairment of peripheral sympathetic transmission found. This is substantiated by recent experiments on dogs reported by Bianchi and Fargier[12] who, during a period of up to 2 hr after i.v. injection of 1 mg/kg reserpine, found no alteration in the catecholamine secretion of the adrenals and no diminution in the catecholamine concentration in the gland. A loss of catecholamine and a lack of responsiveness to nervous stimulation occurs only after 8–24 hr following administration of reserpine.

The diminution in pressor reflexes can thus be taken as evidence of a central site of attack which results in impairment of sympathetic activity. We therefore suggest that reserpine has a dual action on the sympathetic nervous system, i.e. that both central and peripheral sites of attack are involved in the action exerted by reserpine.

REFERENCES

1. Airaksinen, M. M. and Mattila, M. The sedative and lethal actions of reserpine in mice, as modified by 5-hydroxytryptophan, 3,4-dihydroxyphenylalanine, methylphenidate and nikethamide in cold and warm environments. *Acta Pharm. et Toxicol.* **19**, 199–204 (1962).
2. Albrecht, W. Erhöhung der Glykogen-Konzentration im Gehirn und Verhalten verschiedener Fermente in Gehirn und Darm der Maus nach Reserpin (Serpasil). *Klin. Wschr.* **35**, 588–590 (1957).
3. Balzer, H. and Palm, D. Über den Mechanismus der Wirkung des Reserpins auf den Glykogengehalt der Organe. *Arch. Exp. Path. Pharmak.* **243**, 65–84 (1962).

4. Balzer, H., Holtz, P. and Palm, D. Reserpin und γ-Aminobuttersäuregehalt des Gehirns. *Experientia*, **17**, 38 (1961).
5. Baranov, M. N. Der Einfluss von Reserpin auf den Gehalt an Nucleinsäuren in verschiedenen Schichten der Rinde der grossen Hemisphären des Gehirns. *Doklady Akademii Nauk SSSR* **143**, 221—224 (1962).
6. Bein, H. J. Zur Pharmakologie des Reserpin, eines neuen Alkaloids aus *Rauwolfia Serpentina* Benth. *Experientia*, **9**, 107 (1953).
7. Bein, H. J. Significance of selected central mechanisms for the analysis of the action of reserpine. *Ann. N. Y. Acad. Sci.* **61**, 4—16 (1955).
8. Bein, H. J. The action of *Rauwolfia* alkaloids on central mechanisms. Abstracts of reviews. *XXth Int. Physiol. Congress, Brussels* 1956, pp. 455—465.
9. Bein, H. J. The pharmacology of *Rauwolfia. Pharm. Rev.* **8**, 435—483 (1956).
10. Bein, H. J. Effects of reserpine on the functional strata of the nervous system. *Psychotropic Drugs.* Elsevier, Amsterdam, 1957, pp. 325—331.
11. Bein, H. J., Gross, F., Tripod, J. and Meier, R. Experimentelle Untersuchungen über „Serpasil" (Reserpin), ein neues, sehr wirksames Rauwolfiaalkaloid mit neuartiger zentraler Wirkung. *Schweiz. Med. Wschr.* **83**, 1007—1012 (1953).
12. Bianchi, M. and Fargier, M. C. Etude dans le temps de l'effet de la réserpine sur la médullosurrénale. *C. R. Soc. Biol.* **156**, 1797—1802 (1962).
13. Brodie, B. and Beaven, M. A. Neurochemical transducer systems. *Med. Exp.* **8**, 320—351 (1963).
14. Brodie, B. and Shore, P. A. A concept for a role of serotonin and norepinephrine as chemical mediators in the brain. *Ann. N. Y. Acad. Sci.* **66**, 631—642 (1957).
15. Brodie, B, B., Olin, J. S., Kuntzman. R. G. and Shore, P. A. Possible interrelationship between release of brain norepinephrine and serotonin by reserpine. *Science*, **125**, 1293—1294 (1957).
16. Brodie, B. B., Finger, K. F., Orlans, F. B., Quinn, G. P. and Sulser, F. Evidence that tranquilizing action of reserpine is associated with change in brain serotonin and not in brain norepinephrine. Effects of reserpine, raunescine, benzoquinolizines, diethylaminobenzoyl methylreserpate, and stress. *J. Pharmacol. Exper. Therap.* **129**, 250—256 (1960).
17. Callingham, B. A. and Mann, M. Depletion and replacement of the adrenaline and noradrenaline contents of the rat adrenal gland, following treatment with reserpine. *Brit. J. Pharmacol.* **18**, 138—149 (1962).
18. Carlsson, A., Rosengren, E., Bertler, A. and Nilsson, J. Effect of reserpine on the metabolism of catecholamines. *Psychotropic Drugs.* Elsevier Amsterdam, 1957, pp. 363—372.
19. Carlsson, A. and Hillarp, N. A. (1956) quoted after 19.
20. von Euler, U. S. and Lishajko, F. Effect of some drugs on the release of noradrenaline from isolated nerve granules. *Biochem. Pharmacol.* **9**, 77—84 (1962).
21. Freedman, D. X. Psychotomimetic drugs and brain biogenic amines. *Amer. J. Psychiatry*, **119**, 843—850 (1963).
22. Funderburk, W. H., Finger, K. F., Drakontides, A. B. and Schneider, J. A. EEG and biochemical findings with MAO inhibitors. *Ann. N. Y. Acad. Sci.* **96**, 289—302 (1962).
23. Gal, E. M., Drewes, P. A. and Barraclough, C. A. Effect of reserpine and the metabolism of serotonin in tryptophan deficient rats. *Biochem. Pharmacol.* **8**, 32 (1961).
24. Garattini, S. and Valzelli, L. Researches on the mechanism of reserpine sedative action. *Science*, **128**, 1278—1279 (1958).
25. Görög, P. and Szporny, L. Effect of Vincamin on the noradrenaline content of rat tissue. *Biochem. Pharmacol.* **8**, 259—262 (1961).

26. Green, H. and Sawyer, J. L. Intracellular distribution of serotonin in rat brain 1. Effect of reserpine and the monoamine oxidase inhibitors, tranylcypromine and iproniazid. *Arch. Int. Pharmacodyn.* **135**, 426–441 (1962).
27. Heller, A., Harvey, J. A. and Moore, R. Y. A demonstration of a fall in brain serotonin following central nervous system lesions in the rat. *Biochem. Pharmacol.* **11**, 859–866 (1962).
28. Holzbauer, M. and Vogt, M. Depression by reserpine of the noradrenaline concentration in the hypothalamus of the cat. *J. Neurochem.* **1**, 8–11 (1956).
29. Kirpekar, S. M. and Lewis, J. J. Some effects of reserpine and hydralazine upon tissue respiration and the concentration of adenosine nucleotides in certain tissues. *Brit. J. Pharmacol.* **14**, 40–45 (1959).
30. Knoll, J. and Knoll, B. Reserpine: Modification of its tranquilizer effect and analysis of its central mode of action. *Arch. Int. Pharmacodyn.* **133**, 310–326 (1961).
31. Kopin, I. J. and Gordon, E. K. Metabolism of norepinephrine-H^3 released by tyramine and reserpine. *J. Pharmacol. Exper. Therap.* **138**, 351–359 (1962).
32. Kopin, I. J., Hertting, G. and Gordon, E. K. Fate of norepinephrine-H^3 in the isolated perfused rat heart. *J. Pharmacol. Exper. Therap.* **138**, 34–40 (1962).
33. Laverty, R. The influence of cold environmental temperature and histamine treatment on the effect of reserpine in the rat. *J. Neurochem.* **10**, 151–154 (1963).
34. Malhotra, C. L. and Das, P. K. Effect of reserpine on the acetylcholine content of the heart, the ileum and the hypothalamus of the dog. *Brit. J. Pharmacol.* **18**, 190–193 (1962).
35. Mašek, K., Smetana, R. and Rašková, H. Depletion of catecholamines by Shigella Shigae toxin in the mouse brain. *Biochem. Pharmacol.* **8**, 8–9 (1961).
36. Mirkin, B. L. The effect of synaptic blocking agents on reserpine-induced alterations in adrenal medullary and urinary catecholamine levels. *J. Pharmacol. Exper. Therap.* **133**, 34–40 (1961).
36a. Müller, J. M., Schlittler, E. and Bein, H. J. Reserpin, der sedative Wirkstoff aus *Rauwolfia serpentina* Benth. *Experientia*, **8**, 338 (1952).
37. Muscholl, E. and Vogt, M. The action of reserpine on the peripheral sympathetic system. *J. Physiol.* **141**, 132–155 (1958).
38. Olds, J., Killam, K. F. and Bach-y-Rita, P. Self-stimulation of the brain used as a screening method for tranquilizing drugs. *Science*, **124**, 265–266 (1956).
39. Orlans, F. B. and Brodie, B. B. A suggested effect on levels of catecholamine in brain produced by small doses of reserpine. *Nature*, **187**, 1034–1035 (1960).
40. Pepeu, G. and Giarman, N. J. The effect of certain neuropharmacologic agents on brain acetylcholine. *Fed. Proc.* **19**, 280 (1960).
41. Picchioni, A. L., Chin, L. and Breitner, C. Relationship between brain levels of 5-hydroxytryptamine (5-HT) and norepinephrine (NE) and susceptibility to electrically-induced seizures. *Fed. Proc.* **21**, 416 (1962).
42. Pletscher, A., Shore, P. A. and Brodie, B. B. Serotonin release as a possible mechanism of reserpine action. *Science*, **122**, 374–375 (1955).
43. Plummer, A. J., Sheppard, H. and Schulert, A. R. The metabolism of reserpine. In: *Psychotropic Drugs.* Elsevier, Amsterdam, 1957, pp. 350–362.
44. Potter, L. T. and Axelrod, J. Studies on the storage of norepinephrine and the effect of drugs. *J. Pharmacol. Exper. Therap.* **140**, 199–206 (1963).
45. Pscheidt, G. R. and Himwich, H. E. Reserpine, monoamine oxidase inhibitors, and distribution of biogenic amines in monkey brain. *Biochem. Pharmacol.* **12**, 65–71 (1963).
46. Sanan, S. and Vogt, M. Effect of drugs on the noradrenaline content of brain and peripheral tissues and its significance. *Brit. J. Pharmacol.* **18**, 109–127 (1962).

47. SCHANBERG, S. M. and GIARMAN, N. J. Drug-induced alterations in the sub-cellular distribution of 5-hydroxytryptamine in rat's brain. *Biochem. Pharmacol.* **11**, 187–194 (1962).
48. SCHIEVELBEIN, H., WERLE, E. and JACOBY, W. Über die Freisetzung von 5-Hydroxytryptamin (Serotonin) aus Geweben durch Nikotin. *Naturwissenschaften*, **48**, 602–603 (1961).
49. SHEPPARD, H. and ZIMMERMAN, J. H. Reserpine and the levels of serotonin and norepinephrine in the brain. *Nature*, **185**, 40–41 (1960).
50. SHEPPARD, H. and ZIMMERMAN, J. H. A suggested effect on levels of catecholamine in brain produced by small doses of reserpine. *Nature*, **187**, 1034–1035 (1960).
51. SHEPPARD, H., TSIEN, W. H., SIGG, E. B., LUCAS, R. A. and PLUMMER, A. J. The metabolism of reserpine-C^{14}. III. C^{14}-concentration vs. time in the brains and other tissues of rats and guinea-pigs. *Arch. Int. Pharmacodyn.* **113**, 160–168 (1957).
52. SHORE, P. A Release of serotonin and catecholamines by drugs. *Pharm. Rev.* **14**, 531–550 (1926).
53. SPECTOR, S., SHORE, P. A. and BRODIE, B. B. Biochemical and pharmacological effects of the monoamine oxidase inhibitors, iproniazid, 1-phenyl-2-hydrazinopropane (JB 516) and 1-phenyl-3-hydrazinobutane (JB 835). *J. Pharmacol. Exp. Therap.* **128**, 15–21 (1960).
54. STERN, P. and KOCIC-MITROVIC, D. Wirkung des Reserpins auf die Menge der Substanz P im Rattengehirn. *Arch. Exper. Path. Berlin,* **238**, 57–58 (1960).
55. TRENDELENBURG, U. Supersensitivity and subsensitivity to sympathomimetic amines. *Pharm. Rev.* **15**, 225–276 (1963).
56. TRIPOD, J. and MEIER, R. Wirkung von Serpasil (Reserpin, ein neues Alkaloid von *Rauwolfia Serpentina* B.) auf isolierte Kreislauforgane. *Arch. Int. Pharmacodyn.* **97**, 251–266 (1954).
57. VOGT, M. The concentration of sympathin in different parts of the central nervous system under normal conditions and after the administration of drugs. *J. Physiol.* **123**, 451–481 (1954).
58. WALASZEK, E. J., LEICK, J., BUNAG, R. D. and CHAPMAN, J. E. Relationship of serotonin and norepinephrine brain levels to the tranquilizing effect of reserpine in the chicken. *Fed. Proc.* **21**, 323 (1962).
59. WERDINIUS, B. Effect of temperature on the action of reserpine. *Acta Pharmac. Tox. (Kbh.)* **19**, 43–46 (1962).

ELEUTHEROCOCCUS SENTICOSUS – A NEW MEDICINAL HERB OF THE ARALIACEAE FAMILY

I. I. BREKHMAN

Far-Eastern Section of the Siberian Branch of the Academy of Sciences of the U.S.S.R., Vladivostok, U.S.S.R.

OUT OF the many species of the Araliaceae family only *Panax ginseng* has been used in medicine up to now. A comparative pharmacological study of some Far Eastern species of the family made by us has brought out a new valuable medicinal plant *Eleutherococcus senticosus* Maxim. It is a wild shrub 2 -4 m high, growing in abundance in the Far East.

A liquid extract of the root has been tested, which proved to be of slight toxicity. DL_{50} for white mice is equal to 30 ± 6 g/kg. Animals bear well a protracted (1–4 months) administration of the preparation, and the quantity of eaten up food and body weight increases. Increase in body weight is ascertained by experiments on white rats, minks, chickens (Fig. 1) and on white mice and rabbits.

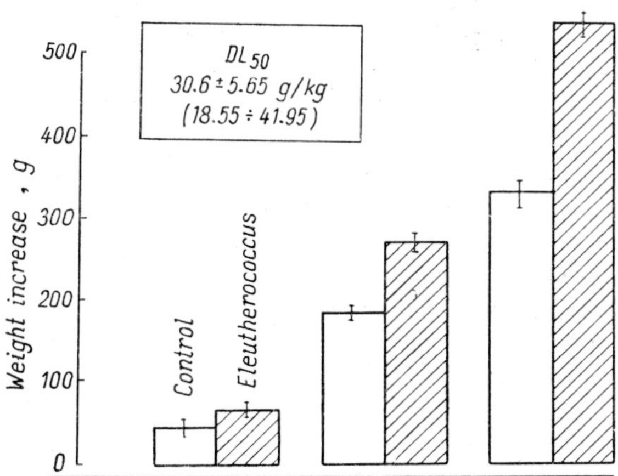

animals	rats		minks		chickens	
days	30		120		100	
N	20	20	59	57	118	112
P	0.009		0.000		0,000	

FIG. 1. Influence of *Eleutherococcus* on body weight of experimental animals.

With a protracted administration of the drug, an increase of glycogen in the liver and muscular tissue as well as of total albumin of blood serum was also ascertained. A still more manifest increase of the albumin of blood

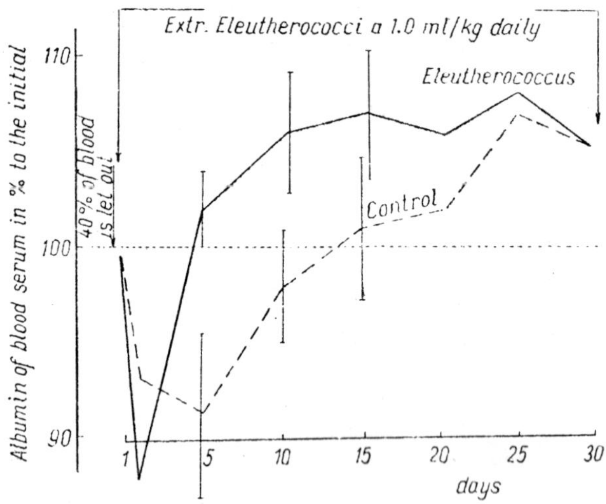

FIG. 2. Influence of *Eleutherococcus* on restoration of blood serum albumin after blood-letting in rabbits.

serum was attained in rabbits after a loss of blood (40 per cent of blood). The original level of the albumin of blood serum was restored on the 15th day in the control group; in that with the preparation it was restored on the 5th day (Fig. 2). These data testify to the anabolic effect.

The gonadotropic effect was established in experiments on immature white mice weighing 7–8 g. During a 30-day period, out of the ten white mice in the control group five were on heat for 1·9 days, whereas in the experiment nine out of ten mice were on heat for 2·9 days. Increase in weight of the seminal vesicles and prostate was recorded in the experiments on males of the same age (Fig. 3).

In Fig. 3 the results of the experiments on 240 minks (females) are also given, half of which received the preparation during all the time of pregnancy. Per hundred females 20 cubs more than in the control group were born. The average weight of a cub was 10 g more on the 21st day and the pre-register mortality was 2·5 per cent less.

An important property of the preparation is increase of labour efficiency and decrease of fatigue. In a special contrivance, mice were made to run up along a descending cord. After a single and to a greater extent a 12-day administration of the preparation the working duration of the mice extended to complete exhaustion (Fig. 4).

Increase of the labour efficiency of man was established with different kinds of dynamic and statical work; indexes of mental work of man improving too (corrective test).

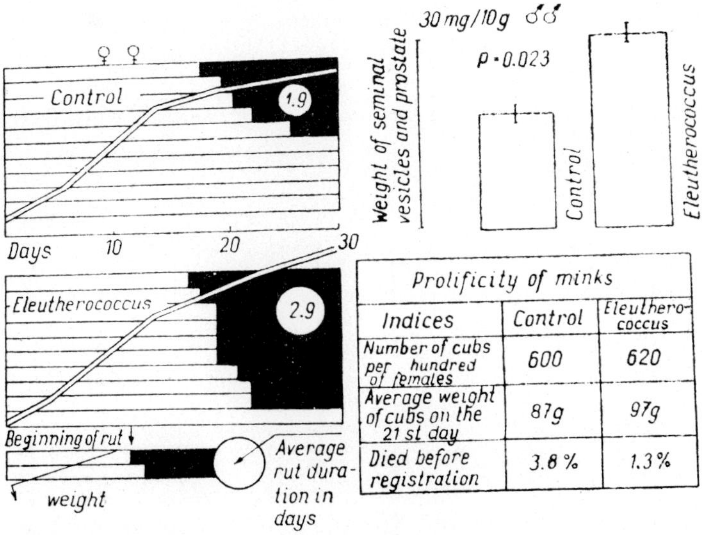

Prolificity of minks

Indices	Control	Eleuthero-coccus
Number of cubs per hundred of females	600	620
Average weight of cubs on the 21 st day	87g	97g
Died before registration	3.8 %	1.3 %

FIG. 3. Effect of *Eleutherococcus* on pubescence of male white mice and the prolificity of minks.

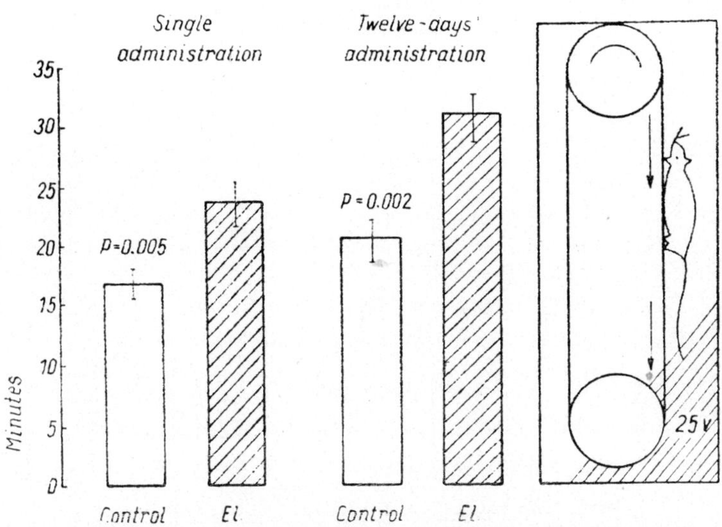

FIG. 4. Duration of work of white mice to complete fatigue after administration of *Eleutherococcus*.

Of great interest is its adaptogenic action — non-specific increase of organism resistance. Data on its adaptogenic action were obtained in experiments on white mice under overload in centrifuge, overheating and cooling (Fig. 5). It was established in experiments on hens that cooling inhibits increase in weight and considerably diminishes egg-laying. Under these conditions a daily 1 ml/kg of the preparation essentially increases resistance in hens. Increase in body weight was steadily going on, and for two cold months the hens laid 2·2 times more eggs than in the control group Fig. 6).

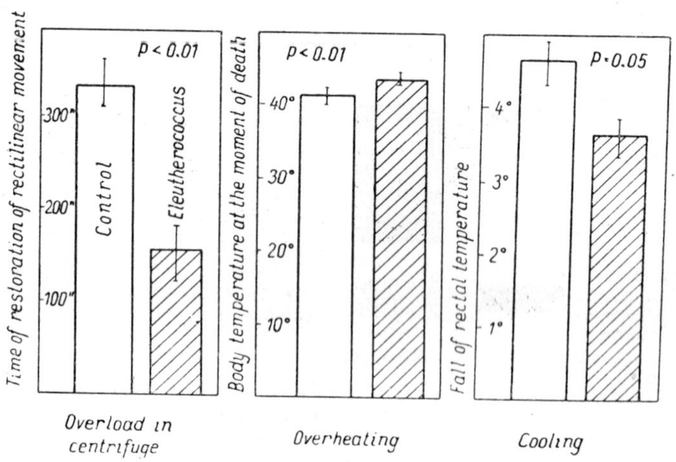

FIG. 5. Adaptogenic effect of *Eleutherococcus* with respect to noxious physical factors in mice.

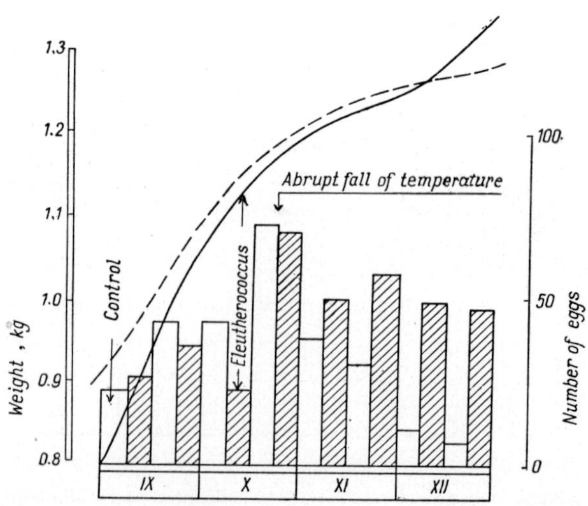

FIG. 6. Influence of *Eleutherococcus* on body weight and egg-laying in hens.

Protective action of the preparation reveals itself towards various noxious factors of physical, chemical and biological nature. It is known that all these factors to a certain extent cause a stress accompanying many pathological states and diseases. In this connection it is worth-while studying the influence of the preparation on stress reactions.

In stress caused by a strong muscular strain in rats immobilized on their backs for 24 hr the preparation lessened the reaction of the adrenal glands (weight, the quantity of ascorbic acid and cholesterol), the involution of the organs of the thymic–lymphatic system (thymus, spleen, lymphatic nodes) and the quantity of stomach ulcerations (Table 1).

TABLE 1

ANTI-STRESS EFFECT OF ELEUTHEROCOCCUS
(immobilization of rats during 24 hr)

Stress symptoms		Control	Stress		
			NaCl 0·85%	*Eleutherococcus*	*P*
Suprarenal	weight, mg/100 g	14·6±0·51	32·1±2·1	25·9±0·9	0·015
	ascorbic acid, mg%/g	354 ±18	209 ±9	260 ±17	0·018
	cholesterol, mg/g	36·8± 1·0	23·9±0·9	26·6±1·2	0·050
Thymus, mg/100 g		42·9± 1·8	25·9±1·6	32·2±2·1	0·022
Spleen, mg/100 g		50·4± 3·9	38·6±3·8	46·1±2·9	0·127
Stomach ulceration	quantity per rat	—	5·8±0·9	2·5±0·9	0·040
	Pauls' index	—	5·22	1·75	—

Protective anti-stress action of *Eleutherococcus* has been established in this and some other models. But stress itself performs a protective function. So it was of interest to elucidate whether a lessening of stress reaction would have any unfavourable influence on the protective function of an organism.

In one experiment rats were made to swim to death. Half of them were killed after 5 hr. The rats in the control group had a greater hypertrophy of the adrenals than those in the experimental group. Of the surviving rats, those under the influence of the preparation were swimming for 10 hr on the average, those in the control group only 9 hr (Fig. 7).

This testifies to the facts that the preparation renders an organism more resistant to stress and to the usefulness of anti-stress action. It has been established too that animals become more resistant to repeating stress action of different nature.

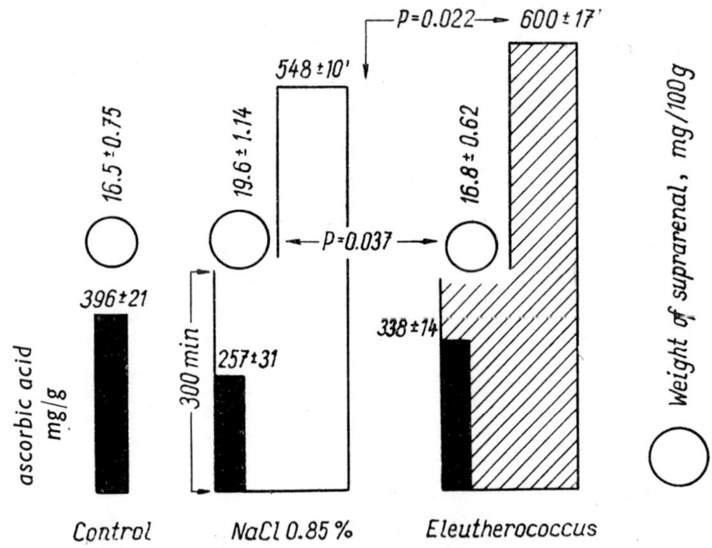

FIG. 7. Influence of *Eleutherococcus* on swimming test, ascorbic acid content and weight of adrenals in rats. White and hatched areas indicate the means of swimming time.

SUMMARY

1. *Eleutherococcus* root is of negligible toxicity and has anabolic, gonadotropic, stimulating and protective adaptogenic action.
2. One of the possible mechanisms of numerous useful effects is the anti-stress action of the preparation.

INDEX